53 Advances in Biochemical Engineering Biotechnology

Managing Editor: A. Fiechter

Springer-Verlag Berlin Heidelberg GmbH

Downstream Processing Biosurfactants Carotenoids

With Contributions by
E. Bárzana, R. Freitag, C. Horváth,
P. K. Roychoudhury, A. Srivastava,
V. Sahai, U. A. Ochsner, T. Hembach,
A. Fiechter, E. A. Johnson, W. A. Schroeder

With 47 Figures and 24 Tables

 Springer

ISBN 978-3-662-14857-0   ISBN 978-3-540-49234-4 (eBook)
DOI 10.1007/978-3-540-49234-4

Library of Congress Catalog Card Number 72-152360

Springer-Verlag Berlin Heidelberg 1996
Originally published by Springer-Verlag Berlin Heidelberg New York in 1996
Softcover reprint of the hardcover 1st edition 1996

Typesetting: Macmillan India Ltd., Bangalore-25
SPIN: 10122587   02/3020 - 5 4 3 2 1 0 - Printed on acid-free paper

Editorial

Professor Fiechter   Professor Scheper

The Hour of Parting

The Series of Advances in Biochemical Engineering/Biotechnology is rooted in a period of scientific changes. Its founding was a response to the implementation of antibiotic production on a large scale. Antibiotic industries emerged after years of struggling for reliable, efficient production processes where aseptic operation on an industrial scale was absolutely necessary. Besides contamination risks the engineers (mostly chemical engineers) were confronted with inadequate process control in aerobic systems with mycelial growht. Oxygen transfer in submerged cultures became a major topic and electronic computing came into use for process modeling. Oxygen transfer, aeration and agitation were the prominent subjects in process improvement.

Secondary metabolites were difficult to produce - the syntheses were low in yield, susceptible to contamination and slow in growth.

Molecular biology also established itself in the 1960s as a new field in the biosciences. As a logical continuation of classical biology, it was an upheaval of an older science which was principally of a descriptive nature. Though still too young for application molecular genetics was soon considered as the future power in pure and applied biology.

No wonder that these fundamental developments were not restricted to the western countries. They soon arrived in the developing nations with their high demand for antibiotics. (Curricula in biotechnology were developed for raising a new generation of engineers interested in the new field of application.)

As in western countries, the definition of biotechnology was not an easy task in a time when the sciences of biology and engineering were considered to be a kind of contradiction. This was not so for Tarun Kumar Ghose at IIT Delhi. He devoted all his energy to introducing of sound and modern teaching courses on all academic levels. Instead of writing a text book, he solicited a number of review articles written by several colleagues from abroad. With this, Volume No. 1 of the Advances in Biochemical Engineering was born. Following our concept, engineering and biologically oriented topics were covered. Process development, bioregulation, downstream processing (but not genetics) normally represented the bio-aspects, whereas modeling of mass transfer, agitation and aeration stood for the engineeringside. Therefore the Series also became quite popular with engineers. The first volume was published in 1971 and earned quite a rewarding and stimulating appreciation. Since then, 52 volumes have been released reporting scientific and technical developments and discoveries. In 1973, the first cloning of a gene was made and used for the biosynthesis of a pharmaceutical product opening new strategies in biosciences and their applications in industry. With this, biology had reached its engineering counterparts now making up the modern biotechnology.

During this same period bioanalytical tools have undergone rapid development. Larger molecules were able to be separated and purified and computer application succeeded in replacing the older devices for process control. On-line analysis has become the basis of automation in bioresearch and process development. Biosensors and dedicated computers were applied in an increasing extent improving reliability, accuracy and precision of the processes. Automation of full R & D programs is now providing new strategies for biological non-invasive research.

The renaming of the Series in 1980 by adding „Biotechnology" to the title was a first response to the ongoing changes indicating that biology has become a hard science and that its application is gradually losing the characteristics of an art and developing toward a quantitative science.

The Advances series will celebrate ist 25th jubilee next year. A new generation of scientists and engineers is now active in the fascinating field of biotechnology. Enthusiastic successors have taken over the task of promoting the modern, molecularly based application of biology.

With this, the hour of parting as Managing Editor has arrived. The Series is in the fortunate position of being directed in future by Prof. Thomas Scheper from the University of Hannover. As a learned biochemist he is one of the promising representatives of the first biotechnology generation whose professional career has been pursued in Institutes where modern biotechnology has been taught. I have so doubt that Advances will flourish under his guidance. My best wishes will accompany him in his interesting and stimulating task.

Zürich, August 1995   Armin Fiechter

Changes

A Note of Thanks and an Announcement of Future Prospects

Professor Armin Fiechter established and, as the managing editor, has maintained the series Advances in Biochemical Engineering/Biotechnology for nearly 25 years. We would like to convey our thanks for all the effort and time he has invested in the series over this period.

He has already pointed out the enormous changes which have taken place in the field of biochemical engineering giving rise to the new discipline of „Biotechnology" and also led to the development of Advances in Biochemical Engineering/Biotechnology.

We are very pleased to have gained Prof. Thomas Scheper as the successor to Prof. Fiechter. He is a scientist who represents the new generation in Biotechnology. As publisher, we are looking forward to co-operating with him and we have no doubt that this co-operation will bring stimulating new ideas to our series.

Heidelberg, September 1995   The Publisher

Attention all "Enzyme Handbook" Users:

A file with the complete volume indexes Vols. 1 through 10 in delimited ASCII format is available for downloading at no charge from the Springer EARN mailbox. Delimited ASCII format can be imported into most databanks.

The file has been compressed using the popular shareware program "PKZIP" (Trademark of PKware INc., PKZIP is available from most BBS and shareware distributors).

This file distributed without any expressed or implied warranty.

To receive this file send an e-mail message to:
SVSERV@DHDSPRI6.BITNET.
The message must be: "GET/ENZHB/ENZ_HB.ZIP".

SPSERV is an automatic data distribution system. It responds to your message. The following commands are available:

HELP	returns a detailed instruction set for the use of SVSERV,
DIR (*name*)	returns a list of files available in the directory "name",
INDEX (*name*)	same as "DIR"
CD <*name*>	changes to directory "name",
SEND <*filename*>	invokes a message with the file "filename"
GET <*filename*>	same as "SEND".

Table of Contents

Gas Phase Biosensors

Eduardo Bárzana
Facultad de Química, Universidad Nacional Autónoma de México,
México D.F. 04510

The last two decades have shown an enormous interest and research efforts on the use of biological activity in analysis, mainly for direct or on line monitoring of chemical compounds. The core work has emphasised on the detection of analytes in aqueous solutions. However, the rapid progress of enzymatic applications on low water environments in recent years is considered to be of high potential for the development of biosensors applicable to compounds present in the gas phase. This review describes the principles that underline the operation of gas phase biosensors and the potential areas of impact in the forthcoming years with examples that illustrate its applicability. A distinct advantage is identified for the direct interaction of gaseous compounds with biological targets that originate a detectable and measurable signal applicable at the field or site of interest, as opposed to preparative steps like the solubilization of volatiles in water, sampling and transportation to the laboratory for further analysis. This is illustrated by enzyme systems based on dehydrated alcohol oxidase for the determination of ethanol or formaldehyde vapors. Current knowledge on the phenomena that control the performance of gas phase enzymatic conversions is also discussed.

The two approaches that have predominated for gas phase biosensors are covered, one based on a biochemical reaction mediated by enzymes or whole cells, and other based on the formation of complexes of biological molecules via adsorption, an alternative in which the use of antibodies has been specially fruitful when coated on piezolectric crystals. The later is unique for environmental applications as in the case of pesticides. Other areas of impact include food related applications and the control of bio-processes.

Advances in Biochemical Engineering/
Biotechnology, Vol. 53
Managing Editor: A. Fiechter
© Springer-Verlag Berlin Heidelberg 1995

1 Introduction

The detection of chemical compounds has been an essential tool for the advancement of chemical and biological sciences since their beginnings. The evolution of modern techniques, like HPLC, GLC, or NMR, with expanded capabilities in terms of precision, reproducibility and sample handling shows the enormous driving force for the sustained progress in analysis.

Within the last two decades an immense interest has been observed in the use of biological activity in analysis. This line of applied research has led to the concept of biosensors, defined as devices or arrangements that detect and measure a variety of compounds based on a molecular modification or interaction brought about by biomolecules with catalytic capabilities. Clearly, biosensors are considered a recent product of biotechnology, since the progress in this multidisciplinary science has resulted in a deeper knowledge of biocatalysis and biological phenomena in general that will translate into innovative possibilities in analysis. In addition, recent discoveries in the fields of electronics, electrochemistry and optical transduction will contribute decisively to the future of biosensors. Consequently, the commercial expectations for biosensors are based on their enormous potential as successors to a wide range of current analytical techniques [1].

The key component of any biosensor is a biological molecule or system, whether in a fairly purified form (enzyme, antibody, nucleic acid, receptor etc.) or as part of a more complex structure (whole cell, organelle, tissue, etc.). The phenomenon involved is the specificity of biomolecules to interact selectively with a particular compound or group of compounds present in a complex mixture, even in extremely low concentrations that can reach the ppb range. Once such a selective recognition event takes place it results in a measurable parameter that has to be amplified to proceed with detection and quantization. The transduction stage is crucial to attain the sensitivity offered by biomolecules and in general can be classified within the following types: amperometric, conductimetric, potentiometric, optical or mass related changes (i.e. vibration frequency in piezolectric crystals) [2].

2 Historical Perspective

So far most research in biosensors has been conducted in aqueous media since the first commercial success of an enzyme electrode developed by Clark and Lyons in 1962 [3]. Applications in clinical chemistry came along that permitted the measurement of components present in biological fluids by means of test kits in a convenient and easy to operate fashion. The independence of laboratory facilities to monitor some key indicators of health status, like glucose in blood or urea in urine, in a timely manner represented a revolution in health services for the general population. Meanwhile, new enzyme assays have been constantly developed for implementation in laboratories and existing ones are being improved, resulting in more rapid and sensitive systems.

A clear indication of the interest and efforts to sustain the high demand for novel systems is represented by a total of 290 research papers related to biosensors published in 1993 alone. It is recognized that recombinant DNA technology is particularly associated with this expansion and a useful tool in the quest for new biosensors [4]. Currently, new fields of application have been identified in which the participation of biosensors would be highly beneficial as in the case of process control in bio-industries or quality control in food operations. However, the greatest impact points to environmental analysis to monitor residuals of man-made chemicals emitted indiscriminately for years that have important consequences in the ecology and the health of individuals.

The uses and applications of biological activity gained new horizons in the early 1980s when Klibanov and coworkers discovered that enzymes in general were able to express their catalytic behavior in fairly pure organic solvents [5–7]. This soon proved to be valid for the construction of enzyme system applicable to the analysis of compounds present in an organic phase [8] opening new possibilities for the determination of analytes with poor solubility in water [9]. As a result, nowadays new strategies on biosensor usage in non-aqueous media have been tested.

Based on the understanding of enzyme behavior in organic media, two new approaches were initiated during the mid 1980s in search of novel biocatalytic activities in environments practically devoid of water; the biocatalysis in supercritical fluids [10, 11] and of compounds present in the gas phase [12]. Both proved to be feasible, proceeding in a reasonable time when careful control of low water content of the system was maintained. The extension of gas phase enzymatic activity to the analysis of gaseous compounds was then attempted and proved to be applicable to the analysis of ethanol and formaldehyde vapors without the need for sample pretreatment procedures [13, 14]. In addition to enzymes, other forms of bioactive molecules that interact with gaseous compounds have been exemplified through the use of antibodies investigated by Guilbault and Luong for volatile organics including pesticides, toxic chemicals and illegal drugs [15].

The routine determination of compounds in the gas phase is a common practice. Some examples include toxic gases ($NH_3, Cl_2, CO, H_2S, HCN$), nerve gases (phosgene), anesthetics (N_2O), flammables (CH_4, propane, ethylene, ethane, etc), organic vapors (hydrocarbons, ethylene oxide, carbon disulfide, mercaptans, formaldehyde), and inorganic vapors (Hg) [14]. The maximum exposure concentration of most toxic gases is at the parts per million level, thereby requiring detection methods with sufficient sensitivity. Portable but expensive gas chromatographs and infrared detectors are currently available for such assays. The enzyme mediated analysis of hazardous gases may represent a simpler and less expensive alternative to these methodologies, while still maintaining the requirements for sensitivity.

Other vapors and gases are present in processing facilities and their measurement can be used as an indication of process performance, leaks, or as a tool for quality control. This may be an important objective as in the case of the food and the bio-industries.

The objective of this paper is to discuss the principles that underline the operation of gas phase biosensors and the potential areas of impact in the forthcoming years with examples that illustrate their applicability. For full development of gas phase biosensors a deeper understanding of the basic phenomena involved is needed. To that end a review of the two lines of research associated will be presented, namely biocatalysis in the gas phase and the adsorption of gases by biomolecules coated in piezoelectric crystals.

3 Bioreaction Based Systems

The analysis of gases by enzymatic methods has been conducted for years in a similar procedure to that used for any soluble compound, namely after the solubilization of the gaseous analyte in an aqueous buffer that contains the transforming enzyme. This was followed by the transduction stage, mostly amperometric or potentiometric, to quantify the analyte. Finally the value in solution was extrapolated to the gas phase concentration using gas-liquid partition coefficients. In a strict sense such systems cannot be considered a gas phase bioassay but a mere adaptation of aqueous enzyme systems. An obvious disadvantage of this type of assay is the preparation of the gas stream in terms of its incorporation into the liquid phase. Clearly an important benefit would be obtained with systems in which the gaseous phase is directly contacted to the bioactive ingredient and a measurable response is amplified or detected in a short time. A distinct advantage of the direct conversion would be its applicability at the field or site of interest without the need for sampling, transportation to the laboratory and other preparative steps.

This approach has been successfully applied to the analysis of ethanol and formaldehyde vapors in simple devices containing dehydrated alcohol oxidase [14]. The development of the technique was oriented to its applicability as an alcohol breath analyzer for legal or personal security purposes. With this methodology it is possible to establish the alcohol content in the blood (BAC) that is directly related to the alcohol content in the breath. The statutory BAC (blood alcohol content) limit in the USA is generally established at $1 \, g \, l^{-1}$ [16]. Products on the market include disposable tubes filled with a strong oxidizing agent, such as $(NH_4)_2Cr_2O_7$ in sulfuric acid impregnated on silica gel, which change color when exposed to ethanol in the breath; however, other organic substances like tobacco smoke or food associated compounds interfere with the reaction.

The system consisted of an enzymatic powder packed in a small glass tube and kept in place by glass wool plugs. The powder was prepared as follows

[14]: a solution containing alcohol oxidase, peroxidase and 2,6 dichloroindophenol (DCIP) was blended with a specific amount of microcrystalline cellulose powder (Avicel) and stirred for a short period. The resulting paste was left to dry under a flow of air until the water content reached a prespecified level (10–60%) and the final powder was stored at 5 °C until needed. The enzymic activity of the solid biocatalyst was determined in aqueous solution. The overall equations describing the proposed enzymatic transformation are as follows:

$$\text{ethanol} + O_2 \xrightarrow{\text{alcohol oxidase}} \text{acetaldehyde} + H_2O_2$$

$$\text{DCIP}_{ox} + H_2O_2 \xrightarrow{\text{peroxidase}} \text{DCIP}_{decomp} + H_2O$$

where DCIP_{ox} and DCIP_{decomp} are the oxidized and decomposed forms of 2,6-dichloroindophenol with distinct blue and colorless colorations respectively. Therefore, decoloration rate was proportional to the amount of ethanol concentration in the gas phase. Kinetic studies in aqueous solution indicated that the rate of reaction was linear with time, dependent on the initial concentration of ethanol and independent of the concentration of color indicator. Hence the time of color change for a fixed ethanol concentration was controlled by manipulating the enzyme to DCIP ratio. This principle was also tested for the determination of ethanol vapors. To that end, the gas phase concentration corresponding to a BAC of $1 \, g \, l^{-1}$ was calculated to be $10 \, \mu g \, l^{-1}$ and the systems proved to be sensitive to such low concentrations with a response time (color change from blue to pale violet) of about only 2.5 min when the water content was adjusted to 35% in the biocatalytic powder [14].

Once the feasibility of using dehydrated enzymes for the semi quantitative determination of gases was demonstrated, it was clearly of interest to see if the principles could be extended to a quantitative technique for use in many other cases such as the determination of toxic gases in emissions or closed environments. The fact that the described system results in variations in absorbance allows for the use of spectrophotometric methods to monitor the reaction. With those changes occurring in a solid matrix, a transmission densitometer was used to monitor color changes. In this case the enzymatic components were dispersed on plastic sheets for thin layer chromatography (TLC) coated with avicel, placed in glass tubes and contacted with ethanol vapors. The complete set-up is schematically presented in Fig. 1 and the results of a typical test are shown in Fig. 2. The results are displayed as the change in absorbance at 605 nm after exposure for a short period of time to varying concentrations of ethanol vapors (expressed as $g \, l^{-1}$) in the aqueous solution equilibrated with air at 37 °C. It is seen that the slope of the plot (or the absorbance at a fixed time) is a function of the alcohol concentration with a good resolution obtained in less than 1 min. This makes the system precise, easy to manipulate and fast in response.

Taking advantage of the fact that the enzyme alcohol oxidase also has the ability to oxidize formaldehyde [17], the quantitative method described was used to test for the presence of formaldehyde in an air stream. As a result, a significant

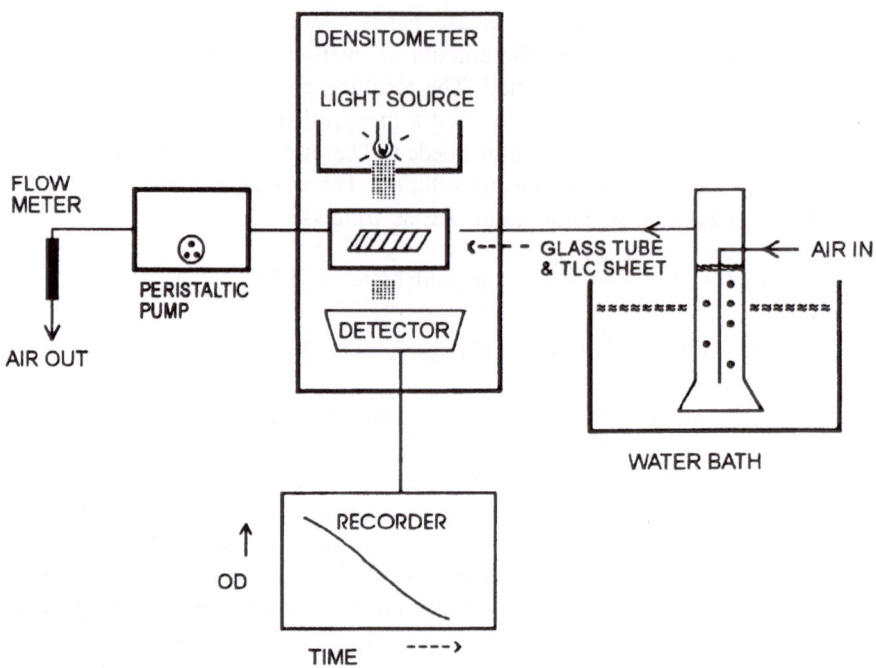

Fig. 1. Schematic representation of quantitative method for the analysis of ethanol and formaldehyde vapors [14]

change in absorbance in less than 1 min was detected indicating that solid alcohol oxidase can also be used for the determination of formaldehyde vapors [14].

Based on the system described, the direct enzymatic or biocatalytic conversion of gases or vapors represents a novel approach not yet fully explored but with high potential in gas phase biosensors. Its development depends on the progress and understanding of the phenomena involved in general for enzymatic reactions in anhydrous media, and in particular for gas phase biocatalysis. A brief review of the state of the art on enzyme related transformations of gaseous substrates is presented in the following section.

The advantages of the direct conversion of gaseous compounds as compared to conventional liquid systems have been described [13, 18] and can be summarized as follows.

– Immobilization is unnecessary since the enzyme has no incentive in abandoning the support. Therefore it can simply be deposited onto the surface of the support to increase the available area of contact. In turn, the biocatalyst can readily be recovered from the reactor.
– The higher difusivities in the gas phase permit a more efficient transfer rate of substrate to the catalytic entity.
– The recovery of products and unconverted substrates is facilitated by simple fractional condensation or adsorption.

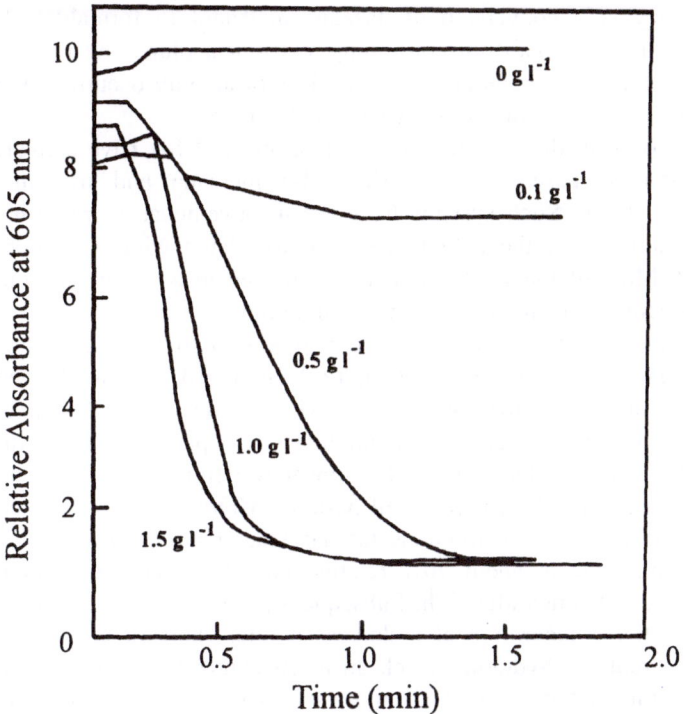

Fig. 2. Response of quantitative method to ethanol vapors. See [14] for details

- The ecological impact is negligible due to the absence of a liquid phase that needs reprocessing or treatment before disposal (aqueous or organic).
- Dehydrated enzymes tend to be more stable to inactivation that their counterparts in solution and are less susceptible to become microbially contaminated.

The historical development of enzyme action on gases was reported by Barzana [13]. The first proof of dry enzyme activity on gaseous substrates was presented by Yagi et al. in Japan [19]. They were able to demonstrate that dry hydrogenase catalyzed the conversion of para-hydrogen to ortho-hydrogen, in addition to a slow hydrogen isotope exchange reaction. The experiments were carried out at room temperature and no conversion was observed when the enzyme had been pre-inactivated at 200 °C for 2 h.

In a second report published 10 years later [20], Yagi et al. reported that, using similar methods and conditions, a dry hydrogenase of high purity also catalyzed the reversible oxidoreduction of cytochrome C3 with molecular hydrogen at 30 °C. For the interconversion reaction, the rate was only 0.22% of that obtained in aqueous solution. The residual water content in the enzyme powder, however, was not determined. In a pioneering study, Cedeño and Waissbluth

[21] reported the enzymatic conversion of gaseous methanol to formaldehyde using freeze dried cells of a methanol consuming yeast in a plug flow reactor. The optimum temperature was above 80 °C and no loss of activity occurred after 1000 h of operation (no temperature is indicated in this case).

The groups of de Bont in the Netherlands and Hou in the USA have actively studied the oxidation of gaseous hydrocarbons by methane and methanol oxidizing bacteria and yeast. Such systems have the inconvenience of requiring the continuous regeneration of the NADH cofactor associated with the mono-oxygenase involved. Most of the work has been conducted in aqueous environments at room temperature using resting cells, immobilized cells or cell free extracts [22–24] although some reports have been produced to test the performance of gas-solid bioreactors. De Bont et al. [25] employed an immobilized cell suspension in a tubular reactor with recycle. The gaseous substrate fed to the reactor was saturated with water. It is mentioned in the paper that "....saturation of the gas phase with water prevents the biocatalyst from drying out and concomitant inactivation". Hou [26] used a cell paste of *Methylosinus sp.* coated on porous glass beads to pack a continuous tubular reactor operated at 40 °C. The catalyst was maintained at about 70% relative humidity and a maximum conversion of 2.7% was obtained after 7 h. Subsequent production occurred only after regeneration of the cofactor but at a decreased rate. Hamstra et al. [27] attempted the use of reduced hydration levels in a gas-solid bioreactor without success due to the rapid loss in oxidizing activity of alginate entrapped cells at water activities below 0.9.

The enzyme alcohol oxidase has been employed extensively by various groups to conduct gas phase bioconversions since the first systematic study on its behavior by Bárzana et al. [12]. It constitutes an appropriate model system from which most of the current understanding of enzyme-gaseous molecule interactions and reactor performance has been obtained. This enzyme has been used either in a purified form or in whole yeast cells or extracts, mostly from *Pichia pastoris*, from which it is commercially obtained. The reaction scheme for alcohol oxidase is the following:

$$\text{alcohol} + \text{oxygen} \longrightarrow \text{aldehyde} + \text{hydrogen peroxide} .$$

The enzyme is active on low molecular weight non-branched alcohols as well as on formaldehyde, albeit at a slower rate [17, 28]. If ethanol is used, acetaldehyde is produced, both compounds with high vapor pressures that volatilize at reasonably low temperatures (30–60 °C). It was demonstrated that a dry preparation of the enzyme adsorbed on DEAE-cellulose vigorously catalyzes the oxidation of ethanol vapors with a strong dependency of the reaction rate on the water content of the active powder [12]. However, the enzymatic activity is severely inhibited by the product hydrogen peroxide that can be prevented by the addition of a decomposing enzyme like catalase or peroxidase. The reaction has been further described in a detailed kinetic study [29] in which the water activity (Aw) was recognized for the first time as a critical parameter, promoting not only the reaction rate but also the rate of catalyst inactivation. For instance, an increase in Aw from 0.11 to 0.97 results in an increase in reaction rate by almost four orders of magnitude, whereas the thermostability decreased by a factor of a hundred as shown in Fig. 3. The same study demonstrated that the formation of a preadsorbed ethanol phase on the surface of the support is not a prerequisite for the reaction, an indication that a direct interaction of gaseous ethanol with the enzyme takes place. Furthermore, at the appropriate hydration level, the reaction rate in the gas phase is comparable to that obtained in aqueous solution. This indicates that the molecular movements needed for catalysis are not restrained in the dehydrated solid preparation.

Alcohol oxidase activity has also been tested successfully in the form of extruded pellets of *Pichia pastoris* cells packed in a continuous flow reactor fed with a mixture of ethanol/water vapors [30]. In addition, acetaldehyde has been produced using non-growing whole cells of *Pichia pastoris* in a semibatch bioreactor with operation time exceeding 100 h with adequate alcohol oxidase activity [31]. Similarly, dried whole cells of *Hansenula polymorpha*, another methanol assimilating yeast, was used to convert ethanol to acetaldehyde in a gas-solid bioreactor with the system conserving its stability over one month at 35 °C, with

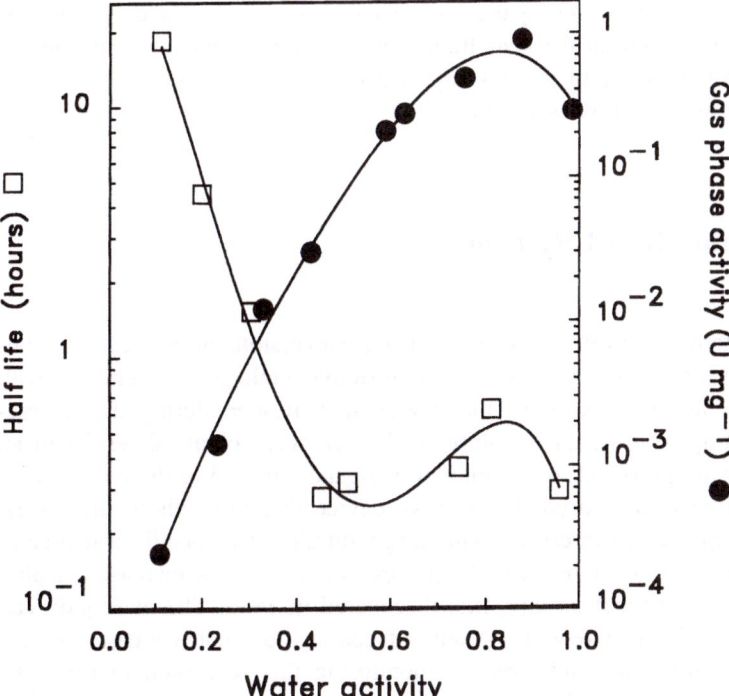

Fig. 3. Effect of water activity on gas phase activity and thermostability of alcohol oxidase acting on ethanol vapors [29]

complete conversion of the substrate, while the water content was maintained below 8% [32].

Immobilized lipases have also been tested successfully in gas-solid bioreactors by Legoy et al. in France [33–37] and by Ross and Schneider [38] in Canada. The reactions studied included some known to be catalyzed by lipases and well characterized in organic solvents: hydrolysis, alcoholysis, ester exchange and esterification. The extent to which a particular reaction predominated depended on water activity values. As expected, hydrolysis predominates at the highest water activities [18]. However, reaction preference between alcoholysis and hydrolysis changed as acyl chain length of substrate increased depending on the enzyme source [38]. Interestingly, the vapor mode of operation over a range of water activities had comparatively small effects on kinetic constants for hydrolysis of ethyl acetate, which were similar to those in phosphate buffer [38].

Thermostable lipases have been used and operated successfully in gas phase conversions up to 100 °C for transesterification reactions [37]. At water contents lower than complete hydration the enzyme shows an adequate balance between stability and activity. Therefore, the combination of water activity and temperature determines the longevity of the enzyme and its activity on gaseous substrates, similar to what occurs with alcohol oxidase [29] and other enzymes in organic solvents.

As mentioned, the biocatalytic conversion of gases is a newly developed area that has received attention during the last seven years. Clearly, a deeper understanding will come along in the near future, opening new applications in bioprocessing and most probably in the development of competitive biosensors for the direct determination of gaseous compounds.

4 Bioadsorption Based Systems

A promising alternative to the enzyme mediated conversion and subsequent analysis of gases is based on the property of bioactive molecules to interact with gaseous ligands as in the case of the associations formed during the interaction antibody-antigen and enzyme-inhibitor. The complex obtained results in an increase in mass proportional to the amount of ligand attached to the biomolecule that can be sensed in an extremely sensitive device that responds to mass variations. In principle, a quartz crystal, vibrating naturally at a specific and precise frequency, shifts its rate of resonant frequency when its mass increases, a phenomenon known as piezoelectricity, first discovered in 1880. The mass gain can be as low as 10^{-12} g or a few Ångstroms of coating to elicit a measurable response [1]. The principles and uses for monitoring the deposition of materials like metals from the vapor phase onto solid substrates have been reported [39]. Therefore this type of arrangement, a piezoelectric crystal coated with bioactive molecules, constitutes an ultra-sensitive microbalance applicable to the detection of molecules with enormous specificity. Similarly, a surface acoustic wave transducer (SAW) can be used at a very high frequency range (more than 9 MHz) resulting in greater sensitivities.

The first analytical use of a piezoelectric crystal was conducted by King [40]. Updated reviews on other uses in biosensors have been published [1, 15]. In 1972 a crystal was coated with an antigen and immersed in a liquid sample to detect the concentration of a specific antibody [41]. Coatings of antibodies or antigen were also tested in similar applications for evaluations in aqueous solutions [42] while Roederer and Baastians [43] dried the crystal after the immunological reaction had taken place and measured the change in resonating frequency in air (proportional to antigen concentration in solution). Also the antibody method was used to assay for the presence of *Candida albincans* in solution, again after drying the crystal [44] with high specificity to this species against others tested.

A dedicated bioassay for the detection of *Salmonella* species was developed by Prusak-Sochaczewski et al. [45] and applied to foodstuffs. The antibody to *S. thyphimurium* was immobilized on a quartz crystal by various procedures including polyethylene-imine layers, glutaraldehyde cross linking, silane layer and through interaction via biotin-avidin. The system responded to a range of 10^5 to 10^9 cfu present in a microbial suspension. The time required for a complete interaction between the crystal and the cells appeared to depend upon the cell concentration of the analyzed sample, and the antibody-bound crystal lost no activity over 4 days at 4 °C while reused for 6–8 consecutive assays.

The promising results on the use of piezoelectric crystals in aqueous solution was extended by Guilbault to assay compounds directly from the gas phase. The first breakthrough on this line of research included the use of formaldehyde dehydrogenase coatings to detect formaldehyde vapors in the range 1–100 ppm [46]. The system was insensitive to potential interfering compounds like other aldehydes or alcohols.

Taking advantage of the fact that many pesticides can be volatilized, the piezoelectric method was adapted to the determination of those toxic compounds in gaseous streams. Based on the demonstrated fact that a wide range of insecticides and some of the nerve gases developed for chemical warfare inhibit the action of cholinesterase in higher organisms, this enzyme was used to coat piezoelectric crystals. The apparatus was sensitive to vapors of various pesticides indicating the formation of an enzyme-inhibitor complex directly from the gas phase in a selective manner, while only minor amounts of water were required for the reaction to proceed [15].

Cholinesterase had been used with success for the evaluation of several pesticides solubilized in water using potentiometric arrangements which require more elaborated experimental manipulations [47]. In those configurations, cholinesterase is normally immobilized in a gel matrix and layered onto a small polyurethane pad. During a test the pad is placed in contact with a solution of the enzyme substrate, butyrylthiocholine iodide, which is readily hydrolysed by the enzyme to form the product thiocholine iodide. When placed in a potentiometric electrode working as transducer, thiocholine iodide is oxidized at the anode, giving rise to a steady potential. However, when a sample containing the inhibitor (i.e. pesticide) is mixed with the substrate solution, the production of thiocholine iodide is reduced and the electrode potential increases in proportion to the concentration of inhibitor present in the sample. It has been reported that the system can be readily adapted for monitoring both air and water samples [47]. Clearly, if good reproducibility and sensitivity can be attained, the use of piezoelectric crystals coated with the enzyme represent a much simpler approach to implement and operate.

When piezoelectric crystals were coated with antibodies raised against parathion, an excellent response was obtained when this pesticide was incorporated into a gas stream at known concentrations [48]. A detection level up to the ppb range was established with a response of 2–3 min and recovery times of 1–2 min. When vapors of other pesticides were contacted with the crystals, some signal was detected but at substantially higher concentrations than those of parathion. Also, nonspecific adsorption on non-antibody protein coatings was not observed for parathion, demonstrating that the identification process performed by the antibody in solution was still active for gaseous molecules. The binding was reversible since the same frequency signal was observed after repeated exposure of the crystal to the antigen. The average lifetime of the coated crystals was about a week, a significantly higher stability than that observed in solution. With the current knowledge on biomolecule stability, this is very probably a result of the low hydration levels of the polypeptide structure.

It has been claimed that this approach could be applicable for detecting the presence of illegal drugs in hidden compartments or shipments [15]. To that end, antibodies to benzoyl ecgonine were used as coating and a good response to cocaine traces in the gas phase was obtained with a high dependency on the relative humidity of the incoming gas stream. The most interesting result is that detection at the part per trillion range is feasible.

The studies discussed in the present paper on the use of piezoelectricity for the fabrication of gas phase biosensors show a very promising potential for practical applications. However, a deeper understanding of molecular interactions in such gas-solid systems is still lacking, restricting usefulness in the analysis of a wider range of organic compounds in gaseous environments.

5 Outlook for Gas Phase Biosensors

Thus far the available reports covered in this review of the use of gas phase biosensor for the direct assay of compounds present in gases and mediated by the action of molecules with biological activity (i.e. enzymes or antibodies) clearly demonstrate the feasibility of this novel approach. In principle, the methods discussed can be adapted to the determination of any chemical compound in the gas phase that can selectively interact with an appropriate enzyme or antibody system producing a change easily detectable by visual observation or instrumental methods. It is estimated that numerous potential applications exist based on the simplicity of their operation with practically no need to pretreat the sample through solubilization in liquids or adsorption, methodologies normally required in other current techniques that rely on more expensive instrumentation.

Gas phase biosensors should have an important participation in diverse areas. As an example of a food related application, we are currently developing a simple biosensor for the detection of methyl amines in the gas phase as an indicator of fish freshness. In this case, the bioanalytical component is the bacterial enzyme trimethylamine dehydrogenase that can be produced biologically [49]. The enzyme degrades trimethylamine via an electronic transfer that can be coupled to oxido-reduction indicators resulting in a detectable color change. Trimethylamine and other amines are generated during fish flesh decomposition through the action of microbial metabolism, resulting in the typical pungent rotten fish odor. Based on this principle, a diagnostic test strip has been developed to establish indirectly fish quality in aqueous suspensions of fish flesh macerates [50]. Whole cells have also been used to monitor trimethylamine in fish using a biosensor that detects the difference in the oxygen uptake of two electrodes coated with *Pseudomonas omniovorans* [51]. In principle, since trimethylamine is a volatile compound, it is susceptible to conversion in the gas phase by dehydrated preparations of the dehydrogenase in a system analogous to the one reported by Bárzana et al. [14]. Other aromas, gaseous products of food degradation, volatile precursors, etc. could be determined using this strategy provided that specific enzymes are available.

Other areas of interest for gas phase biosensors is the control of bioprocesses, since many gases and vapors are associated with the metabolic status of the producing microorganism. The importance of continuous measurement has been addressed [52]. Traditionally, mass spectrometry, a high cost alternative, has been routinely used for on line detection of exit gases that can be translated into environmental adjustments or corrective actions on process variables. The field represents an excellent opportunity for tailored gas phase biosensors.

Probably, the greatest potential for biosensors in general, and for gas phase detection in particular, is the contribution they may have to the chemical characterization of the environment [47]. Since most pollutants are biodegradable and harmful for only a short period until they are broken down by the natural degrading routes of the environment, it is in theory possible to manufacture the enzymatic system specific for each pollutant by selection and propagation of the appropriate living host. This is also applicable to compounds of high resistance to degradation, provided that an appropriate screening and selection methodology is conducted. In fact, many biological degrading systems for xenobiotics have been produced with high activity levels by genetic engineering manipulations.

In the atmosphere of populated urban areas and in confined areas like underground parking lots, mine shafts, ship decks, offices, etc., many compounds may become health hazards when present above specific threshold concentration values. Examples include carbon monoxide, hydrogen sulfide and even gaseous metals like mercury, for which a biological method could be developed. Other gaseous compounds are constantly discovered to be present at undesirably high levels and represent a fertile field of application for biosensors. For instance, formaldehyde is an extremely toxic volatile chemical implicated not long ago as a carcinogenic agent and the US Occupational Safety and Health Administration has set a maximum exposure limit of 1 ppm gaseous formaldehyde over an 8-h period [53]. It is slowly released from composites fabricated with formaldehyde based resins commonly used in furniture and construction panels. A high level has been found in mobile houses. Within the same concept, many household products like air fresheners, pest controllers and the like are probably overused in many instances, representing a serious health hazard with long term effects. Constant monitoring of the responsible chemical compounds using simple in-house devices would be highly recommended, representing an enormous market for gas phase biosensors.

The future of gas phase biosensors will depend in many respects on the expense associated with their production as compared to other analytical methodologies. Therefore, the availability of biomolecules with enhanced activities at low cost is critical. In addition, this activity has to be protected from adverse modifications that occur during the preparation of such biomolecules in the form of a dehydrated active powder included in the biosensor. In this regard, little is known about the mechanisms that result in activity lost for most enzymes during drying, storage, further hydration to an active state and while in operation. Fortunately, the area of stability of dry proteins has gained renewed attention in the last years, given its importance in the preparation of biocatalysts effective in organic media. Some recent reports clearly show this tendency [54, 55]. It is then expected that the number of related studies will grow significantly in the near future to guarantee eventually the accessibility of highly potent preparations; this will certainly open new opportunities for gas phase bioassays and bioconversions.

6 References

1. Luong JHT, Mulchandani AM, Guilbault GG (1988) TIBTECH 6: 310–316
2. Hall EAH (1992) Overview of biosensors. In: Edelman PG, Wang J (eds) Biosensors and chemical sensors. ACS Symposium Series 487, American Chemical Society, Washington, DC
3. Clark LC, Lyons C (1962) Ann NY Acad Sci 102: 29–45
4. Brodelius PE (1991) Curr Op Biotechnol 2: 23–29
5. Zaks A, Klibanov AM (1984) Science 224: 1249–1251
6. Zaks A, Klibanov AM (1985) Proc Natl Acad Sci USA, 82: 3192–3196
7. Klibanov AM (1986) Chemtech 16: 354–359
8. Kazandjian RZ, Dordick JS, Klibanov AM (1986) Biotechnol Bioeng 28: 417–421
9. Saini S, Hall GF, Downs MEA, Turner APF (1991) Anal Chim Acta 249: 1–15
10. Hammond DA, Karel M, Klibanov AM, Krukonis VJ (1985) Appl Biochem Biotechnol 11: 393–400
11. Randolph TW, Blanch HW, Prausnitz JM, Wilke CR (1985) Biotechnol Lett 7: 325–328
12. Barzana E, Klibanov A, Karel M (1987) Appl Biochem Biotechnol 15: 25–34

13. Barzana E, PhD Thesis, Massachusetts Institute of Technology, Cambridge, USA
14. Barzana E, Klibanov AM, Karel M (1989) Anal Biochem 182: 109–115
15. Guilbault GG, Luong JH (1988) J Biotechnol 9: 1–10
16. Flores A, Eliason LK, Wu YC (1981) Nat Bur Stand Special Publication 480–41, US Department of Commerce, Washington, DC
17. Patel RN, Hou CT, Laskin AI, Derelanko P (1981) Arch Biochem Biophys 210: 481–488
18. Lamare S, Legoy MD (1993) TIBTECH 11: 413–418
19. Yagi T, Tsuda M, Mori Y, Inokuchi H (1969) J Am Chem Soc 91: 2801
20. Kimura K, Suzuki A, Inokichi H, Yagi T (1979) Biochim Biophys Acta 567: 96–105
21. Cedeño M, Waissbluth M (1978) Enzyme Engineering, Brown GB, Maneckee G, Wingard LB (eds) Vol. 4, pp 405–407. Plenum, New York
22. Hou CT, Patel R, Laskin AI, Barnable N (1979) Appl Environ Microbiol 38: 127–134
23. Hou CT (1982) US Patent 4: 348,476
24. Habets-Crutzen AGH, Brink LES, van Ginkel CG, de Bont JAM, Tramper J (1984) Appl Microbiol Biotechnol 20: 245–250
25. de Bont JAM, van Ginkel CG, Tramper J, Luyben KChAM (1983) Enzyme Microb Technol 5: 55–59
26. Hou CT (1984) Appl Microbiol Biotechnol 19: 1–4
27. Hamstra RS, Murris MR, Tramper J (1987) Biotechnol Bioeng 29: 884–891
28. Hopkins TR, Muller F (1987) In: (van Verseveld HW, Duine JA, eds) Microbial growth on C1 compounds pp 150–157, Nijhoff, Dordrecht, The Netherlands
29. Barzana E, Karel M, Klibanov AM (1989) Biotechnol Bioeng 34: 1178–1185
30. Duff SJB, Murray WD (1990) Process Biochem 25: 40–42
31. Chiang HK, Foutch GL, Fish WW (1991) Appl Biochem Biotechnol 28–9: 513–525
32. Kim CH, Rhee SK (1992) Biotechnol Lett 14: 1059–1064
33. Pulvin S, Legoy MD, Lortie R, Pensa M, Thomas D (1986) Biotechnol Lett 8: 783–784
34. Pulvin S, Parvaresh F, Thomas D, Legoy MD (1988) Ann NY Acad Sci 613: 303–312
35. Parvaresh F, Vic G, Thomas D, Legoy MD (1990) Ann NY Acad Sci 613: 303–312
36. Parvaresh F, Robert H, Thomas D, Legoy MD (1992) Biotechnol Bioeng 39: 467–473
37. Robert H, Lamare S, Parvaresh F, Legoy MD (1992) Prog Biotechnol 8: 85–92
38. Ross NW, Schneider H (1991) Enzyme Microb Technol 13: 370–377
39. Hillman AR, Loveday DC, Swaan MJ, Bruckenstein S, Wilde CP (1992) Analytical applications of the electrochemical quartz crystal microbalance. In: Edelman PG, Wang J Biosensors and chemical sensors. ACS Symposium Series 487, American Chemical Society, Washington
40. King WH, Jr (1964) Anal Chem 36: 1735–1739
41. Shons A, Dorman F, Najarian JJ (1972) J Biomed Mater Res 6: 565–570
42. Rice TK (1980) US Patent 4: 314, 821, Feb 9, 1982
43. Roederer JE, Baastians GJ (1983) Anal Chem 55: 2333–2336
44. Karube I, Gotoch M (1987) In: Guilbault GG, Mascini M (eds) Analytical uses of immobilized biological compounds for detection, medical and industrial uses, Reidel Pub Co
45. Prusak-Sochaczewski E, Luong JHT, Luong, Guilbault GG (1990) Enzyme Microb Technol 12: 173–177
46. Guilbault GG (1984) Anal Chem 55: 1682–1684
47. Bickerstaff GF (1987) Enzymes in industry and medicine. New Studies in Biology Series. Edward Arnold, London
48. Ngeh-Ngwainbi J, Foley PH, Kuan SS, Guilbault GG (1986) J Am Chem Soc 108: 5444–5447
49. MacIntire WS (1990) Methods Enzymol 188: 250–260
50. Wong K, Gill TA (1988) J Food Sci 53: 1653
51. Gamati S, Luong JHT, Mulchandani A (1991) Biosensors & Bioelectronics (1991) 6: 125–131
52. Mandenius CF, Danielsson B, Mattiasson B (1984) Anal Chim Acta 163: 3–15
53. Marshall E (1987) Science 236: 381
54. Liu WR, Langer R, Klibanov AM (1991) Biotechnol Bioeng 37: 177–184
55. Roziewski K, Russell AJ (1992) Biotechnol Bioeng 39: 1171–1175

Chromatography in the Downstream Processing of Biotechnological Products

Ruth Freitag[1] and Csaba Horváth[2]
[1]Institut für Technische Chemie, Universität Hannover, Callinstr. 3,
D-30167 Hannover, Germany
[2]Department of Chemical Engineering, Yale University, Mason Laboratory,
New Haven, CT 06520, USA

Chromatographic techniques are essential for the isolation and purification of most of the high value products of modern biotechnology. The economically sensible and technically satisfactory downstream processing of a therapeutic protein, usually involves a number of chromatographic steps. Its development and optimization require considerable knowledge of the various physico-chemical and engineering aspects of biochemical chromatography. This review addresses the various modes of chromatography and the design of chromatographic separation processes from a biotechnologist's point of view. Strategies for optimizing the structure of the downstream process are outlined and scaling up considerations are discussed. The importance of the different chromatographic methods in research and development is estimated in an analysis of protein purification schemes recently published in the literature. Finally, examples of the application of chromatographic procedures for process scale product purification in the biotechnological industry are given.

Advances in Biochemical Engineering/
Biotechnology, Vol. 53
Managing Editor: A. Fiechter
© Springer-Verlag Berlin Heidelberg 1995

List of Abbreviations

BSA	Bovine serum albumin
CHO	Chinese hamster ovary
CIP	Clean in place
CM	Carboxymethyl
DEAE	Diethylaminoethyl
DNA	Desoxyribonucleic acid
EDTA	Ethylendiamine tetraactetic acid
FA	Fluoroapatite
FPLC	Fast protein liquid chromatography
GMP	Good manufacturing practice
GPC	Gel permeation chromatography
HA	Hydroxyapatite
hGH	Human growth hormon
HIC	Hydrophobic interaction chromatography
HPLC	High performance liquid chromatography
HSA	Human serum albumin
IAS	Ideal adsorbed solution
IDA	Iminodiacetic acid
IEC	Ion exchange chromatography
IGF	Insulin like growth factor
IgG	Immunoglobuline G
IMAC	Immobilized metal affinity chromatography
mAb	Monoclonal antibody
MIC	Metal interaction chromatography
PEI	Polyethyleneimide
RPC	Reversed phase chromatography
SEC	Size exclusion chromatography
SIP	Sanitize in place
SOP	Standard operating procedure
TNF	Tumor necrosis factor
TPA	Tissue plasminogen activator

List of Symbols

A	Constant
B	Constant
c	Concentration in the mobile phase
c_n	Displacer concentration
D	Effective dispersion coefficient
d_p	Particle diameter

h	Reduced plate height
H	Height of a theoretical plate
$H(t)$	Step function
k'	Retention factor
K_D	Distribution ratio
L	Column length
q	Concentration in the stationary phase
R	Resolution
t	Time
t_0	Hold up time of an inert tracer
t_r	Retention (hold up) time of a retained substance
u_0	Flow velocity
V_i	Intraparticular void volume
V_0	Interstitial volume
V_r	Retention volume
w	Average peak width measured by the base line intercepts
$w_{0.5}$	Peak width at half height
z	Distance in the direction of the bulk flow
α	Interstitial porosity
$\Delta(z)$	Distance between the maxima of two peaks
$\delta(t)$	Dirac function
ε	Intraparticulate column porosity
θ	Tortuosity
ϕ	Phase ratio
ν	Reduced velocity
τ	Time of sample introduction

1 Introduction

Biotechnology has brought about the controlled and specific industrial exploitation of biological organisms and processes. The scope of the field is increasing steadily due to the continuing progress in recombinant DNA technology. Biotechnological products come in many forms and from many sources. They may be metabolites of microorganisms, higher plant or animal cells, stem from an enzymatic reaction or peptide synthesis, or simply be found in a natural source such as blood plasma, cow milk or plant material. If, for example, the source is a microorganism, the desired substance may be excreted in the culture medium, be enriched in the cyto- or periplasma or form inclusion bodies. Product concentrations may vary from several hundred $g\,l^{-1}$, as in the case of ethanol or citric acid, down to $\mu g\,l^{-1}$ or even $pg\,l^{-1}$, as in the case of blood factors and therapeutic enzymes.

The set of complex isolation and purification steps involved in the recovery of the product is generally referred to as downstream process. In the design of the process leading to product recovery the nature of the starting material is a key parameter. Equally important, however, is the desired quality of the final product, i.e., the maximum acceptable level and the chemical nature of impurities or contaminants. The demands regarding purity will be highest for human therapeutics, somewhat less for a number of diagnostics and often considerably lower for industrial enzymes. Regulatory issues and environmental considerations (GMP, various national laws, etc.) are also important, especially in the case of industrial downstream processing. The design and validation of standard operating procedures (SOP), between batch in-place cleaning (CIP) and sanitizing (SIP) procedures or the validation of operational parameters such as column leeching and the removal of certain impurities (endotoxins, virus DNA, etc.) have to be taken into account.

While the isolation/purification of a highly concentrated, low value product such as ethanol is relatively straightforward and based on standard operations in chemical engineering, the downstream processing of biotechnological products such as Factor VIII [1], monoclonal antibodies [2, 3], or vaccines [4], is much more involved [5–8]. In general the contribution of downstream processing to the overall production cost depends on the upstream product concentration [9–11]. It may vary from less than 10% in the case of citric acid to more than 90% in the case of some of the above mentioned high value products. As a result a direct relationship has been observed between the selling price and the product concentration in the original feed (Fig. 1, [9, 12]).

A typical recovery process can be roughly subdivided into four steps [13–17]. First, the product is separated from the producing organisms and other insolubles by a solid/liquid separation step, such as centrifugation or filtration. This may require cell rupture in the case of intracellularly enriched substances or

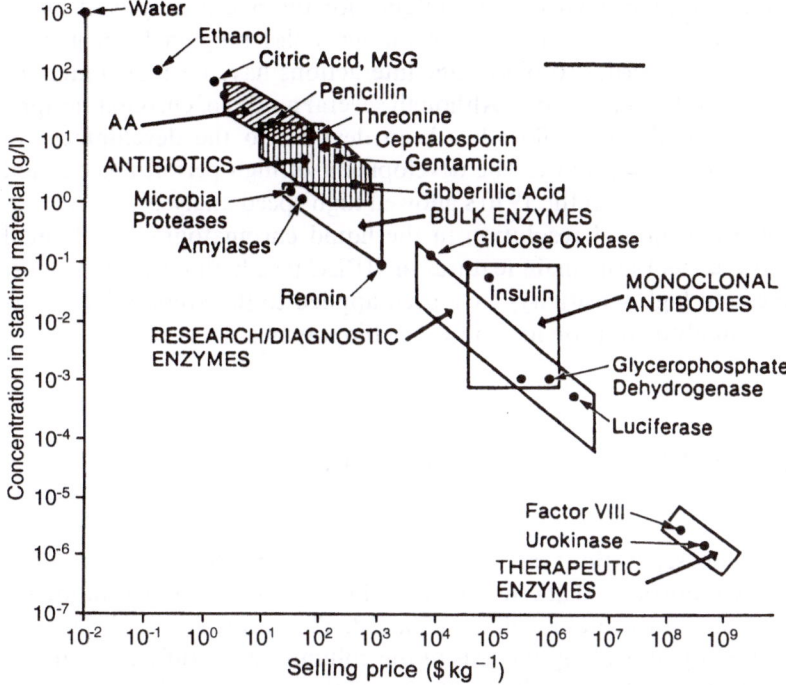

Fig. 1. Relationship between the product concentration in the starting material and the selling price (reproduced with permission from Ref. 9)

resolubilization in the case of inclusion bodies [18]. In the second isolation step, substances that differ considerably in their physico-chemical character are removed from the product. The methods used are either highly specific and based on bioaffinity interactions or nonspecific such as salting out, solvent extraction or batch adsorption. If well designed, this step should result in a considerable increase in product concentration to facilitate the following purification steps. These usually employ highly selective methods to remove from the product substances having similar physical properties and biochemical functions. At this point, chromatography plays a predominant role as a recent analysis of more than a hundred laboratory-scale protein purifications has demonstrated [19]. The fourth step is polishing and may include gel filtration, crystallization, and lyophilization of the final product.

Chromatography has been the primary preparative separation method in biology and biochemistry since 1906, when the Italo-Russian botanist Tswett separated the pigments of chlorophyll by passing petroleum ether extracts through a column packed with powdered chalk [20]. In the 1930s, column chromatography became an important tool for the separation of natural products in general, while the introduction of paper chromatography pioneered the use of chromatographic separation techniques for the microanalysis of biological samples. Various characteristics of biomolecules ranging from general physico-chemical properties to biospecific interactions have been exploited for their chromatographic separations. Although several modes of chromatography have been recognized, most effort has been devoted to the development of (linear) elution chromatography. The development of high performance liquid chromatography (HPLC) in the 1970's allowed high speed analysis and set new standards of precision and resolution in the liquid chromatography of small molecules. In the 1980's the main features of HPLC (i.e., high resolution, short analysis time, and high sensitivity) have been applied to the preparative separation and fast analysis of proteins and other biopolymers.

2 Principles of Chromatographic Separations

Chromatographic separations are traditionally semi-batch procedures. The sample mixture is introduced at one end of a column packed with the stationary phase through which the mobile phase is passed. The separation of the sample components takes places along the axis of the column due to differences in the molecular dimensions or due to hydrophobic, ionic, or biospecific interactions between the sample molecules and the stationary phase. On the production scale, a continuous separation process would be preferable, as such processes are usually more economical, can be automated more easily, and achieve a more standardized product quality. Furthermore, they enable a better utilization of the adsorbant and mobile phase and may facilitate recycling. The advantages of continuous chromatographic separations are fully realized only in large scale processes; therefore such chromatographic processes are found mainly in the petrochemical and sugar industries. However, as industrial biotechnology is advancing, continuous chromatographic systems are expected to play an increasing role also for product isolation and purification.

2.1 Size-Exclusion Chromatography

In size-exclusion chromatography (SEC), also called gel filtration or gel permeation chromatography (GPC), the sample molecules are separated according to their hydrodynamic diameter [21–23]. The sample is passed through a column packed with an inert porous material that possesses appropriate pore size distribution and volume. Separation occurs due to differences in the intraparticulate void volume explored by the sample components of different molecular dimensions. Molecules larger than the upper exclusion limit cannot enter the intraparticulate void space, whereas sufficiently small molecules have access to all the pores. For sample components of intermediate molecular dimensions, the retention volume, V_r, is given by:

$$V_r = V_0 + K_D V_i \qquad (1)$$

where V_0 is the interstitial volume, K_D is the distribution ratio and V_i is the intraparticulate void volume.

The magnitude of K_D is determined by that fraction of intraparticulate volume which can be entered by the sample component under consideration. Plots of K_D versus the logarithm of the molecular weight are generally linear between K_D values of approximately 0.15 and 0.8, and SEC may thus also be used for a rough estimation of the molecular weight of macromolecules [24]. Several models have been put forward to relate K_D to the molecule and porous matrix properties [25–27]. In biochemical applications, SEC is most suitable for rapid and convenient separation of sample components having substantially different molecular weights such as in the desalting of a protein solution. SEC is often used as a polishing step following other chromatographic separations [28, 29]. In analytical SEC, the sample size should be no more than 3% of the column volume; however, in preparative work the sample may occupy up to 15% of the column volume.

SEC is commonly practiced with crosslinked dextran such as Sephadex from Pharmacia, modified agarose and polyacrylamide-based gels that do not generally permit the use of high pressures. However, column packings of high mechanical stability, ranging from silica-based materials to macroreticular rigid polymers, are increasingly employed in SEC-HPLC.

2.2 Adsorption Chromatography

Adsorption chromatography is based on the different distribution of each substance between the stationary and the mobile phase due to interactions between the components and the chromatographic surface. The process can be mathematically described by solving the mass balance equation with the appropriate boundary conditions for a particular chromatographic system [30]. The one dimensional mass balance equation for component i is given as:

$$\frac{dc_i}{d_t} + \phi \frac{dq_i}{dt} + u_0 \frac{dc_i}{dz} = D \frac{d^2 c_i}{dz^2} \qquad i = 1, 2, ..., n \qquad (2)$$

where c_i and q_i are the respective concentrations of i in the mobile and stationary phases, u_0 is the flow velocity of the mobile phase, z is the distance in the direction of the bulk flow, t is time and ϕ is the phase ratio of the column for the components under consideration. The effective dispersion coefficient, D, lumps together all contributions to axial dispersion, and n is the number of components in the system. In order to solve Eq. (2), two boundary and one initial condition must be specified.

The initial condition is generally given by:

$$c_i(0, z) = 0; \quad 0 \leq z \leq L \qquad (3a)$$

and the exit boundary condition by:

$$\left(\frac{dc_i}{dz}\right)_{z=L} = 0 \tag{3b}$$

where L is the length of the column.

According to the inlet boundary conditions, three modes of chromatography can be distinguished, as first described by Tiselius [31]:

Elution $\qquad\qquad\qquad c_i(t,0) = c_{0,i} \quad 0 < t \leq \tau$ \hfill (4a)

$\qquad$ (analytical) $\quad c_i(t,0) = c_{0,i}\,\delta(t)$

Frontal $\qquad\qquad\qquad c_i(t,0) = c_{0,i}\,H(t)$ \hfill (4b)

Displacement $\qquad\qquad c_i(t,0) = c_{0,i} \quad 0 < t \leq \tau \quad i = 1, 2, \ldots, n-1$ \hfill (4c)

$\qquad\qquad\qquad\qquad\quad c_n(t,0) = c_{0,n}\,H(t-\tau)$

where $H(t)$ is the unit step function, $\delta(t)$ is the Dirac delta function, τ is the time of sample introduction and c_n is the diplacer concentration. The mass balance equations has been solved for different systems by using these initial and boundary conditions [32–36].

The equilibrium relationship for the distribution of the components between the mobile and stationary phases is given by the respective adsorption isotherms. In the case of biochromatography, most of the experimental single component adsorption isotherms recorded can be described over a fairly wide concentration range by the Langmuir model [37]. At low sample concentrations, Henry's law usually holds and the relationship between c_i and q_i is linear. Analytical elution chromatography is carried out in that range. At higher concentrations, the isotherm becomes nonlinear. Consequently, competition between the various molecules for the adsorption sites takes place and the component isotherms will be suppressed by the other substances. While multicomponent Langmuirian adsorption isotherms can be constructed using the parameters of the corresponding single component isotherms, this approach is not thermodynamically consistent in most cases [38] and usually does not adequately describe the adsorption behavior of complex molecules over the wide concentration range encountered in preparative separations of biopolymers. Moreover, biopolymers have a tendency for multipoint adsorption, agglomeration and multiple retention mechanisms that complicate their chromatographic behavior. Depending on the individual sample composition and concentration, they may be prone to inter- and intramolecular interactions. The slow diffusion rate of large molecules frequently prevents the attainment of true adsorption equilibrium within the practical time scale of chromatographic separations.

The direct determination of multicomponent isotherms is experimentally rather involved [39–43]. On the other hand the use of single component isotherm parameters in the multicomponent Langmuir isotherm formalism can be misleading and does not provide for selectivity reversal, for instance [44].

Other models, such as the multivalent ion-exchange model and the ideal adsorbed solution (IAS) method, have been suggested to describe the multicomponent adsorption behavior of complex molecules under linear and non-linear chromatographic conditions [45–50].

3 Process Design in Chromatography

3.1 Semibatch Chromatography

3.1.1 Elution Chromatography

As mentioned above, linear elution chromatography is the preferred operational mode in analytical chromatography. The sample is introduced into the column approximately as a Dirac pulse. If the composition of the mobile phase does not change throughout the separation, the process is referred to as isocratic elution. The components traverse the column with different velocities, due to differences in their distribution between the mobile and the stationary phases, and are thus separated. For each component the dimensionless retention factor k', which represents the ratio of the amounts of a given component on the stationary phase to the amounts in the mobile phase, is used to measure the magnitude of the interaction with the stationary phase, i.e., the retention. In the case of isocratic linear elution chromatography at constant flow rate the retention factor of a component is given by:

$$k' = (V_r - V_0)/V_0 = (t_r - t_0)/t_0 \tag{5}$$

where V_r is the retention volume and t_r is the retention time of the sample component and V_0 and t_0 are the hold-up volume and hold-up time of the mobile phase which are usually measured by an inert tracer.

Depending on their k' values, the substances move down the column as individual zones with different velocities that are lower than that of the mobile phase. The degree of separation is expressed by the resolution, that depends on two factors: differences in the retention times and the extent of band spreading. The resolution R is defined as:

$$R = \Delta Z/w \tag{6}$$

where ΔZ is the distance between the maxima of the two peaks on the chromatogram and w is the average peak width as measured by the base line intercepts. An R value of unity represents 98% separation and is considered adequate.

As a result of longitudinal diffusion, mass transport resistances and flow maldistribution, the peak width increases as the sample molecules move through the column. The most common measure of column efficiency is the plate height, H, which has been related to various column parameters and operational variables [51–57]. In order to compare band spreading over a wide range of conditions, it is convenient to use dimensionless quantities. The reduced plate height, h, is given by H/d_p, and the reduced flow velocity v is given by $u_0 d_p/D$, where d_p is the particle diameter of the column packing. In the chemical engineering literature, v is also called Peclet Number, Pe.

In the absence of kinetic resistances for adsorption the reduced plate height can be expressed as the sum of three plate height increments, each dependent on the flow velocity [58, 59]:

$$h = h_{\text{disp}} + h_{\text{ext}} + h_{\text{int}} \tag{7}$$

The first term on the right hand side, h_{disp}, expresses the magnitude of band spreading by axial dispersion due to longitudinal diffusion and maldistribution of flow. It can be written as:

$$h_{\text{disp}} = A + B/v \tag{8}$$

where A and B are constants for a given set of conditions [56].

For an unretained solute, the plate height increment representing external mass transfer resistances, h_{ext}, is given by [59–61]:

$$h_{\text{ext}} = \frac{(1 - \alpha)\varepsilon^2 \alpha^{5/3}}{3.27[\alpha + (1 - \alpha)\varepsilon]^2} v^{2/3} \tag{9}$$

where ε is the intraparticulate column porosity, or the ratio of the intraparticulate volume to the volume occupied by the stationary phase particles, and is the interstitial column porosity, i.e., the ratio of the interparticulate volume to the total column volume. An alternative calculation for this particular term is for example, presented, in Ref. 56.

The plate height increment h_{int} represents the mass transfer resistance due to intraparticulate diffusion, and for an unretained solute is given by [56, 58, 59]:

$$h_{\text{int}} = \frac{\theta\alpha(1 - \alpha)\varepsilon}{30[\alpha + (1 - \alpha)\varepsilon]^2} v \tag{10}$$

where θ is the tortuousity.

For any retained component, the magnitude of both h_{ext} and h_{int} is also dependent on the retention factor. It is assumed in these equations that the intraparticulate mass transport is solely by diffusion. For gigaporous HPLC phases, where this is not the case, the treatment presented in Ref. 61–63 is suggested.

In order to study column behavior, it is customary to plot the plate height against the flow velocity and this is called a van Deemter or Knox plot [51, 52]. In chromatographic practice, H, is usually evaluated as

$$H = \left(\frac{w_{0.5}}{t_0}\right)^2 \frac{L}{5.545} \tag{11}$$

where $w_{0.5}$ is the peak width at half height.

For conventional porous stationary phases the van Deemter plot runs through a minimum. At low flow rates, molecular diffusion is the predominant reason for zone spreading. At comparatively high flow rates of the mobile phase, (intraparticulate) mass transfer resistance and non-equilibria become responsible. The minimum value of H should be between $2d_p$ and $3d_p$ for a well-packed column and the corresponding flow velocity is termed optimum velocity. The resolving power of a column is highest when operated at the optimum reduced flow velocity. Due to the low effective diffusion coefficient of high molecular weight molecules, this will often be impractical for biopolymers.

Protein adsorption tends to follow the all-or-nothing principle [64], i.e., save for a narrow window of eluent strength, the molecules will either adsorb strongly to the surface of the stationary phase or show no tendency to adsorb at all. Since the elution windows of all sample proteins will rarely overlap, their separation by isocratic elution is difficult. Instead the eluent strength is increased gradually or stepwise (differential elution) during the chromatographic run, to bring about the separation of a multicomponent protein mixture. Differential elution is commonly used in preparative/process chromatography. Ideally two well-chosen elution steps will suffice, one for recovering all components that bind less strongly to the chromatographic surface than the product and the other eluting the product exclusively. If the eluent strength is increased gradually, the migration velocity of the eluates increases with the eluent strength, until they move with the velocity of the mobile phase. Since the eluate molecules at the rear of a peak are moving in a zone of higher eluent strength, they are sped up in relation to the bulk sample, while the opposite effect is operative at the front of each sample zone. Therefore elution with an appropriate gradient involves focussing; this results in relatively sharp peaks and a reduction of peak "tailing". At the same time, gradient elution is much more demanding than isocratic elution as far as instrumentation and theoretical treatment of the process are concerned.

Linear elution chromatography does not utilize the full capacity of either the stationary or the mobile phase in the column. Whenever the column is overloaded the sample concentrations are in the non-linear range of the respective adsorption isotherms. Under such conditions the sample molecules compete for the binding sites on the chromatographic surface and interfere with each other's migration. The value of the retention factor as well as the shape of the sample zones strongly depend on the type of the adsorption isotherms and on the component concentrations (Fig. 2, [44]). In the case of a single-component Langmuirian isotherm, the zone will, with increasing substance concentration, increasingly have a sharp front and a diffuse rear boundary or "tail" due to the self-interference of the molecules. The retention factor decreases with increasing "overloading", i.e. sample concentration or volume.

A simple procedure to establish the load in overloaded elution chromatography is touching-band optimization [65], where the sample load is increased until the front of the second component zone just touches the rear of the first. If the column is severely overloaded, separation becomes largely a matter of the sample composition. In certain cases, the competition of the two components will cause the zone of the less strongly bound component to be pushed ahead by

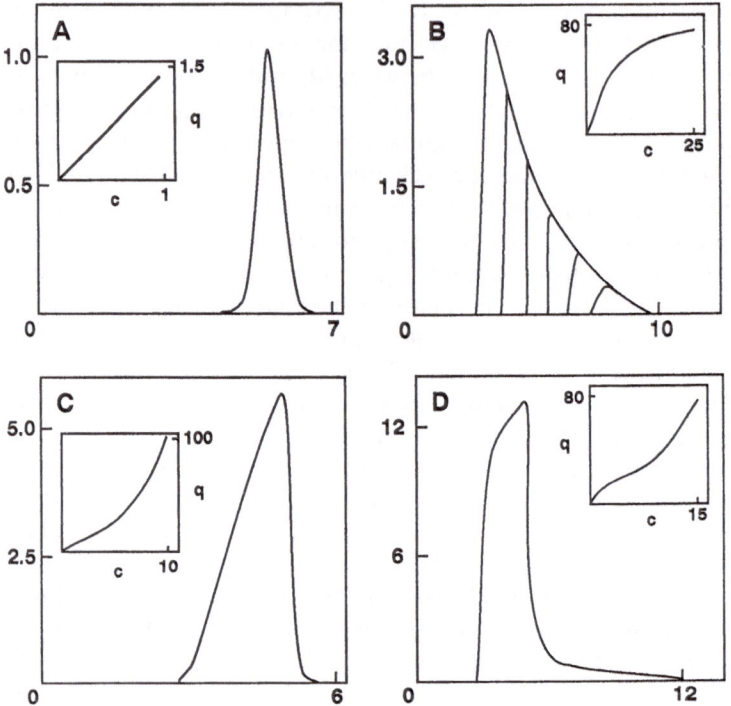

Fig. 2A–D. Relationship between the shape of the single component isotherm and the eluting substance zone (reproduced with permission from Ref. 44)

the zone of the more strongly bound one; "sample displacement" occurs, often with beneficial effects on the separation. The opposite, a "tag-along" effect, may also be observed [66]. The importance of column efficiency is often ignored in preparative chromatography under non-linear conditions. Yet efficient columns have their advantages not only in analytical but also in preparative work because they offer greater throughput and higher purity than columns having lower efficiency. Therefore high performance stationary phases are of advantage in preparative chromatography if high purity of the product is required.

If only two components or fractions need to be separated, recycle chromatography may be considered in nonlinear production scale purifications [67]. In conventional adsorption chromatography, the length of the column is selected so that a complete separation of the components is accomplished. In recycle chromatography, a shorter column length is employed. Therefore, from a binary feed three fractions are collected: pure A, a mixture of A and B and pure B. The mixture, which may comprise up to 40% of the total volume collected, is added to the fresh feed in the following run. For each column length and feed composition, there is an optimal ratio. Too little fresh feed decreases the product concentration and increases the eluent consumption, whereas too much decreases the productivity. The advantages of the recycling method stem from the possibility of using shorter columns, which not only means faster separations and the recovery of more highly concentrated pure A and B but also lower consumption of stationary and mobile phases. This more than outweighs the disadvantage of having to rechromatograph a part of each feed.

3.1.2 Frontal Chromatography

Frontal chromatography constitutes a binary separation process in which only the least-retained component can be separated from the others. The mixture to be separated is fed continuously into the column under conditions that favor the binding of all the components but one. This component is obtained in pure form at the column outlet, until the dynamic capacity of the stationary phase is exhausted and the other sample components break through, Fig. 3. Frontal chromatography is the first step in many biopolymer purification schemes involving differential elution. The method per se is applicable when the product to be purified has much lower affinity for the stationary phase than the other feed components and therefore breaks through far ahead of the impurities.

3.1.3 Displacement Chromatography

This non-linear multicomponent separation technique is eminently suitable for preparative/process scale applications. In displacement chromatography, the competition of the feed components at the chromatographic surface is exploited to bring about the separation of the sample components. Although the principles of displacement chromatography have been known for more than 50 years, the advent of highly efficient HPLC instruments and columns, together with an improved understanding of the theory of nonlinear chromatography

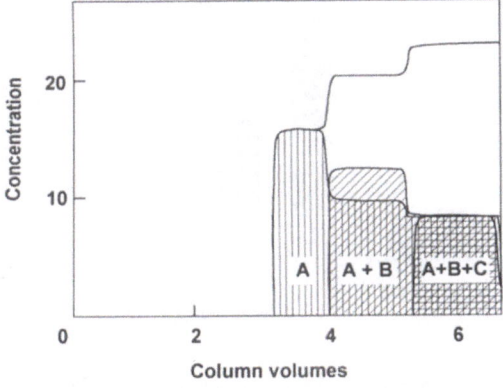

Fig. 3. Separation of a ternary (1:1:1) mixture by frontal chromatography. The corresponding single component isotherms are assumed to follow the Langmuir equation (reproduced with permission from Ref. 34)

have recently provided new impetus for the employment of this chromatographic mode [68–74].

In displacement chromatography (Fig. 4) the feed is loaded onto the column
under conditions that allow strong binding of all sample components to the
stationary phase. Following the feed the displacer, a substance with extremely
high affinity for the stationary phase, is introduced into the column. As the
displacer front traverses the column the components are forced to compete for
the adsorption sites and – at least for systems showing approximately Lang-
muirian-type isotherms – are finally separated into adjacent rectangular bands,
if the column is sufficiently long. At this point all bands move with the speed of
the displacer front and the scaled isotachic state ("displacement train") is
reached. In the "displacement train" the concentration of each zone is determined by the multicomponent isotherms of the respective substances as well as
by the isotherm and concentration of the displacer ([68, 75], Fig. 5). The
concentration of the components may thus be increased with respect to their
concentration in the feed. This feature is of interest not only in preparative scale
separations, but also for the enrichment of certain components in trace analyses
[76, 77]. The effect of feed and displacer concentrations, bed length and mass-
transfer effects on the separation have been treated theoretically and in many
cases the results have experimental support [35, 78–85].

The problems associated with displacement chromatography originate from
the non-linear nature of the process, the paucity of isotherm data and the
concomitant use of the Langmuir isotherm model by many researchers in the
field. For ion-exchange displacement chromatography the steric mass action
model appears to yield good results [49]. Isotherm crossing, resulting in a con-
centration-dependent reversal of selectivity, can be a particularly serious im-
pediment [41, 44, 86]. It requires changes in the operating conditions or even the
use of another column in order to carry out the separation. Selecting the proper
displacer that meets all the requirements can also present a problem particularly

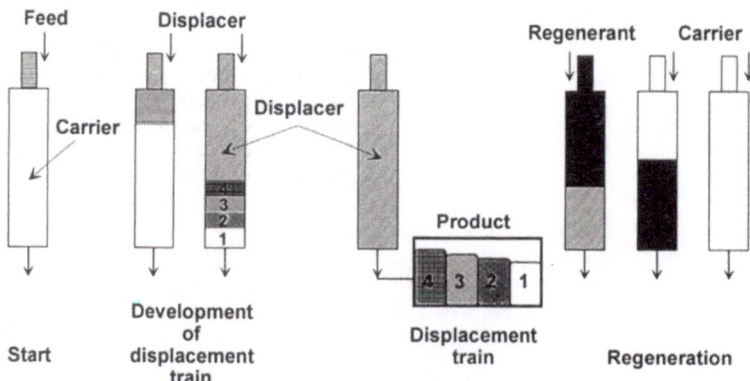

Fig. 4. Stages of displacement chromatography (reproduced with permission from Ref. 78)

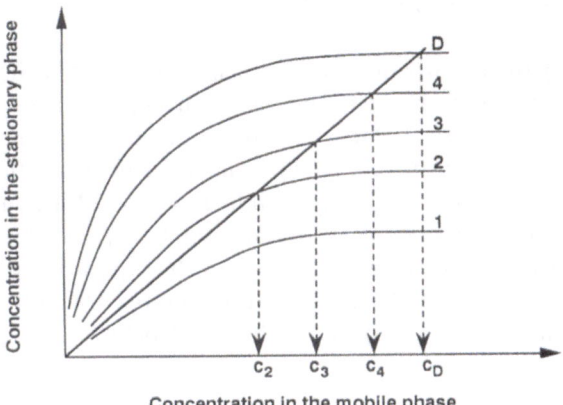

Fig. 5. Displacement chromatography: multicomponent isotherm-system of the feed components and the displacer. The corresponding fully developed displacement train is shown below. The concentration in the substance zones depends on the location of the intersection point between the respective substance's isotherm and the operating line, i.e. the displacer concentration. The volume of the substance zone then depends on the concentration in the feed (reproduced with permission from Ref. 78)

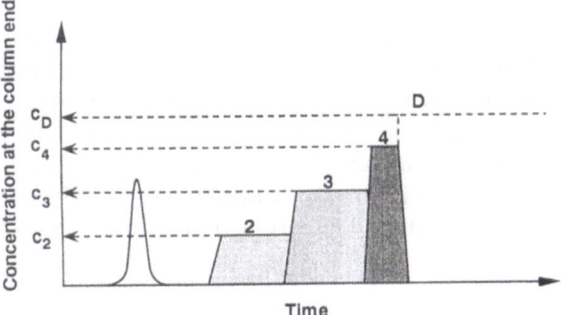

when biopolymers have to be separated. Although further work is needed to exploit the potential of displacement chromatography, results accumulated over the last 14 years have demonstrated that the technique can be a powerful tool for the purification of antibiotics, peptides and even proteins [87–92]. Displacement chromatography has been used to separate the antibiotic Cephalosporin C from culture supernatant [93], to isolate alkaline phosphatase enriched in the periplasm of *E. coli* [94], as well as to isolate IgG from plasma [95] and monoclonal antibodies from ascites [96]. Guinea pig serum proteins and mouse liver cytosol proteins have been isolated by anion-exchange displacement chromatography [97]. Recently, recombinant human antithrombin III has been purified from culture supernatants of Chinese Hamster Ovary (CHO) cells [98]. In analytical biotechnology tryptic digests were characterized by the tandem use of high performance displacement chromatography and mass spectrometry [99]. A comparison of overloaded elution and displacement in preparative chromatography showed that the choice between the two modes will have to depend largely on the comparative economics of displacer recovery and fraction concentrations [100].

3.2 Continuous Chromatographic Separators

In continuous chromatography the adsorbent and the eluent move in opposite directions with respect to the point of sample introduction. Such a system can be organized with a moving bed, moving column, or the simulated moving bed/column (moving port) approach ([101], Fig. 6). The movement of the adsorptive solid and the flow of the eluent may be perpendicular (cross-flow), or opposite (counter-current). While multicomponent mixtures can be separated in cross-flow systems, counter-current systems are restricted to the separation of two component mixtures or the separation of a given mixture into two fractions. Some approaches to continuous chromatography are discussed below.

3.2.1 Continuous Rotating Annular Chromatography

The concept of this technique has been put forward some time ago [102, 103]. The packed bed of the adsorbent occupies the annular space between two coaxial cylinders. The bed rotates past a fixed port through which sample is continuously fed. The eluent percolates downwards through the bed. The sample molecules are separated due to differences in the ratio of vertical to circumferential movement and can be withdrawn continuously at different points at the bottom of the annular column. The stronger the adsorption of

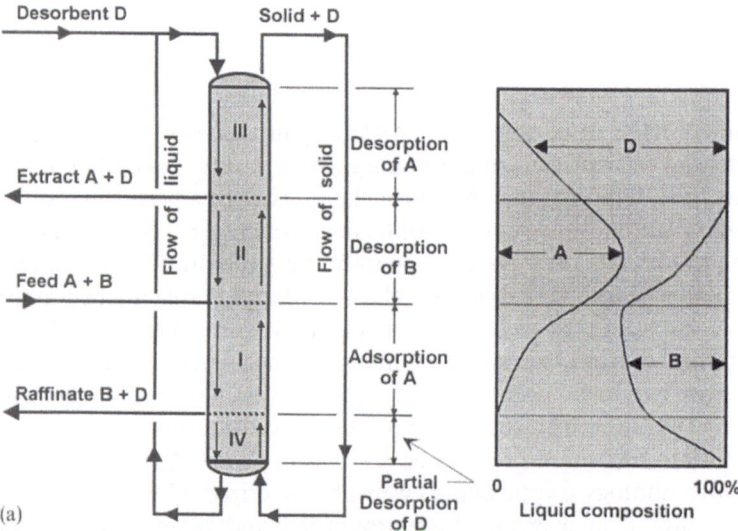

Fig. 6a–f. Types of continuous chromatographic separators: (a) Adsorptive separation with moving bed, (b) Fluidized bed, (c) Magnetically stabilized fluidized bed, (d) Rotating annular chromatograph, (e) Simulated moving bed/column design (Sequential operation of SCCR systems), (f) Sorbex process (reproduced with permission from Refs. 124 (a, e), 111 (b), 119 (c, d), 125 (f))

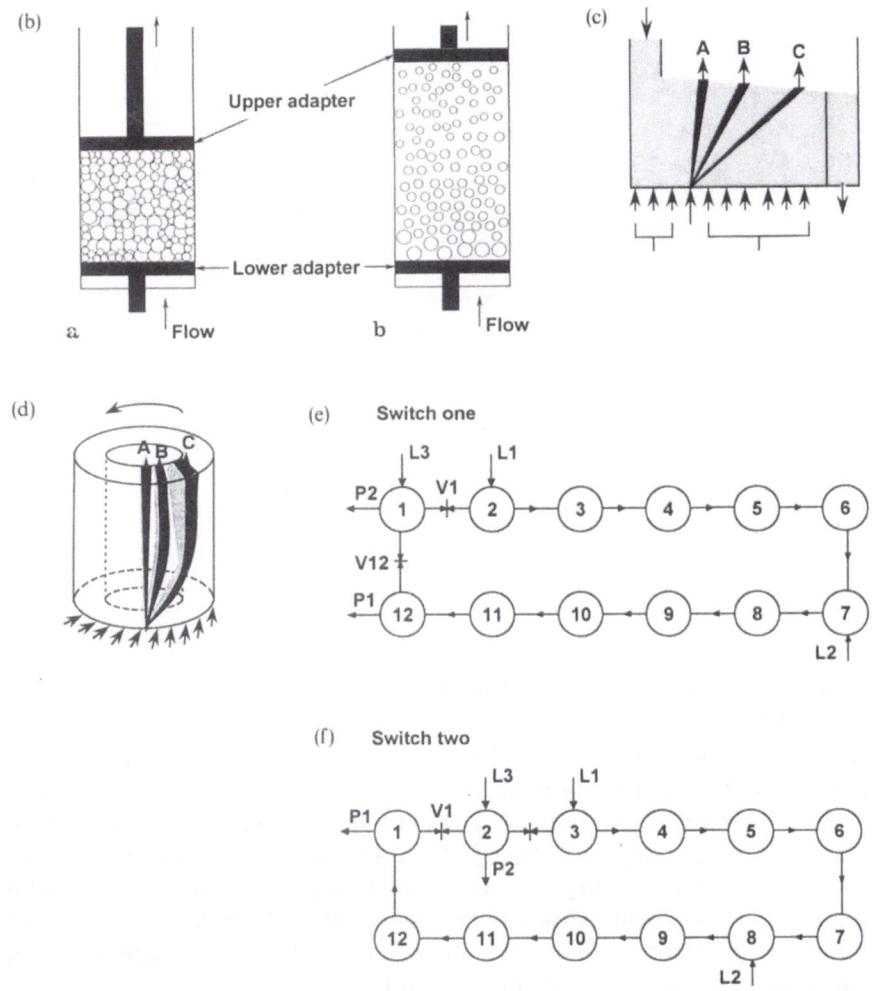

a given substance is, the further away from the point of sample introduction it will appear at the bottom of the bed. The use of gradient elution [104, 105], and that of a pressurized annular chromatograph [106] have been reported. Cow heart myoglobin, skim milk proteins and amino acids have been separated using such systems [101]. Since actually rotating the bed is mechanically difficult, systems where the sample inlet collecting ports are moved in a coordinated manner in relation to the static bed (simulated moving bed) have also been used [101].

3.2.2 Fluidized Bed Adsorption

In a fluidized bed the adsorbent particles are placed in a vessel having a porous bottom. A fluid flows upwards through the porous bottom at a flow rate such

(f)

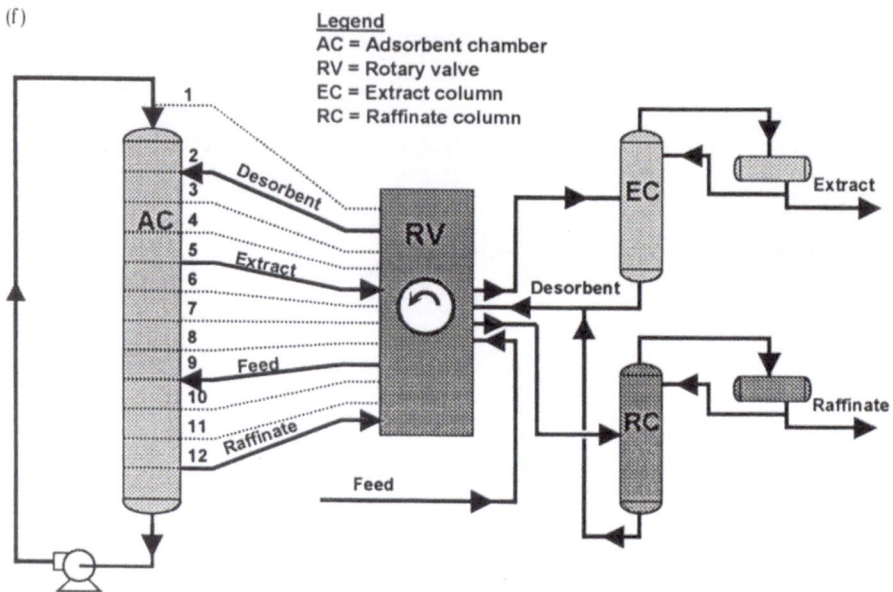

that the particles become "fluidized" within the confines of the container. There are several ways to carry out adsorptive separations with fluidized beds in biotechnology. One application called "expanded chromatographic beds", is employed early in a purification scheme, as it offers the possibility of removing solids while simultaneously concentrating and to a certain extent purifying the product in a single step operation [107–111]. The solids move unhindered through the bed and are removed with the fluidizing liquid, while the product is bound on the adsorbent particles. The use of two fluidized beds has been described for the Continuous Affinity Recycle Extraction (CARE), process: one for adsorption and the other for desorption [112–114].

Since the 1970's, the stabilization by magnetic field of a fluidized bed of magnetic particles has been investigated [115, 116]. This was shown to suppress particle circulation to a large extent and solids seem to move in nearly plug-flow manner in such fluidized beds. As both the solid and the fluid movement are controllable in such systems, magnetically stabilized fluidized beds are interesting as mechanically complex continuous chromatographic separators. Both cross-flow fluidized beds, where the solids move perpendicular to the direction of the fluid and counter current fluidized beds have been described [117–119]. Some examples involving such beds are the separation of lysozyme and myoglobin by ion-exchange chromatography [119] and human serum albumin by Cibacron Blue affinity chromatography [120]. Also, trypsin was isolated in particles carrying trypsin inhibitor ligands [121]. Since solids can be exchanged and recycled in fluidized beds and clogging is not a serious problem, such beds are putatively interesting systems for biotechnological applications.

3.2.3 Simulated Moving Bed/Column Systems

In simulated moving bed (SMB) systems [122, 123] such as the Sorbex, Molex and Olex processes developed at Universal Oil Products (UOP), USA [124] or the semicontinuous chromatographic refiner (SCCR) developed at Aston University, UK [125], a number of columns are connected so that by using an appropriate valving system the operation of a counter-current system is simulated. Such systems have already been used for the purification of dextrans, carbohydrates and sugars using size-exclusion and ion-exchange chromatography [125–128]. Recently, high performance simulated moving bed adsorbers have been developed for industrial use [125]. Whereas such systems are not capable of resolving multicomponent mixtures, they offer a promising high throughput process for binary separations.

4 Adsorbers for Biopolymer Chromatography

Traditionally, stationary phases based on neutral polysaccharides such as cellulose, cross-linked dextran and agarose or on polyacrylamide have been used in the liquid chromatography of biopolymers [129]. Since such materials have generally poor mechanical stability, the separation has to be carried out at low pressures and consequently at low mobile phase flow rates. The lack of mechanical stability impedes scaling up the process with soft gel packings because of the limitations on the column length. While such soft gels continue to be popular largely due to their excellent biocompatibility, a number of more efficient high performance adsorbents have have been introduced recently in biopolymer chromatography. During the last ten years Fast Protein Liquid Chromatography (FPLC) developed by Pharmacia, Uppsala, Sweden to meet the specific demands of biopolymer separation has been the most widely used method. The stationary phases used in such systems are based on macroreticular styrenic support and are stable in a pH range from 2 to 13. Due to the employment of the 10 μm monodisperse beads and comparatively short columns, the column inlet pressure does not need to be so high in order to obtain a relatively high flow. Thus, biocompatible materials such as glass and plastics can be used for wetted parts. Prepacked columns of up to 20 ml are available. The use of even larger columns has been occasionally reported [130].

Composite stationary phases consisting of a rigid, porous microparticulate support and a covalently bound layer with appropriate interactive groups on the surface have been developed for biopolymer separation [131–134]. Originally having been developed for the fast analytical separations of small molecules, they are increasingly used in biopolymer HPLC. The small diameter (2 μm to 10 μm) and narrow particle size distribution of such materials yield high column efficiency but also require high pressures in order to reach a reasonable flow rate of the mobile phase through these columns. In most HPLC units pressures up to 400 bars can be adjusted and such systems usually have to be made with stainless steel wetted parts. As a rule, proteins and most other biopolymers are compatible with stainless steel and most substances of interest can conceivably be isolated in a standard HPLC system without damage. However, adventitious metal ions may affect not only the properties of some biological substances but also the control of the chromatographic process.

Conventional silica supports cannot withstand contact with alkaline solutions; hence this precludes column cleaning with caustic solutions, which is often required in protein chromatography. On the other hand, various stationary phases based on polymers such as highly cross-linked agarose, agarose/dextran, polyacrylamide and polyacrylamide/dextran have been introduced [135–141]. In addition to having much better mechanical properties than the traditional polysaccharide gels (i.e. being almost as rigid as the silica based ones), polymer-based particles are stable in aqueous solutions over a wide pH range and are consequently suitable for biopolymer chromatography at relatively high flow velocities.

In order to have a high chromatographic surface area and thus an acceptable column loading capacity in preparative chromatography, porous stationary phases are used. With an average pore size of 100–300 Å, the mass transport within the particle pores, where most of the adsorptive surface is located, is mainly diffusive in such conventional materials. In the case of macromolecular substances, the slowness of diffusive mass transfer in the pores gives rise to considerable band-broadening and poor resolution [57, 142, 143]. In analytical biopolymer chromatography, micropellicular nonporous stationary phases offer high speed and high column efficiency [138, 144–146], but they are less suitable for preparative applications due to the somewhat lower capacity of such columns. Recently, gigaporous stationary phases have been introduced for rapid separation of biopolymers [147–154]. Such particles have a bidispersive pore structure: large through-pores (gigapores) of several hundred μm diameter, which reach across the particles, and small diffusive pores, at the walls of the gigapores to increase the adsorptive surface. Thus only small distances have to be covered by diffusion [150]. Under appropriate conditions convective mixing occurs in the gigapores with a concomitant enhancement of intraparticulate mass transfer. As a result, adsorption equilibria are approached more rapidly and much higher flow rates can be used. The use of these stationary phases is referred to as perfusion chromatography [151–154]. Resolution and capacity seem to be almost independent of the flow rate of the mobile phase [61, 147, 154].

In recent years, several stationary phase systems based on membrane configurations have been introduced. Such membrane adsorbers (MA) (i.e. filter membranes that have been functionalized by the attachment of interactive groups) might offer certain advantages in biopolymer chromatography [155–161]. The thickness of the functionalized MA ranges from 100 μm to several mm. Mass transfer to the adsorptive surface takes place by convection rather than by diffusion [162–164], but lateral mixing due to anastomosis is poor and therefore the efficiency is limited. The low back-pressure caused by the MA allows the use of extremely high flow rates [165–167]. Since the separation efficiency (plate height) shows hardly any dependence on the flow rate, separations can be carried out within seconds at mobile phase flow rates of 50 ml min^{-1} and more, unless the adsorption kinetics themselves constitute the limiting step [167, 176]. The suitability for high flow rates constitutes an obvious advantage in the downstream processing of the typical dilute biotechnological feed solution. Moreover, whereas the scale-up is difficult with particulate chromatographic columns (see below), this is hardly a problem in the case of MA-chromatography, where the expertise gained in standard filtration processes might be used to advantage.

To date, MA-chromatography is employed mainly in affinity-based separations: plasma proteins [168], antibodies [169], recombinant interleucin [170, 171], tissue plasmin activator [172], fibrinogen and IgG [173,174], and microbial enzymes [175] have e.g. been isolated by so-called specific filtration. In this context, metal affinity (interaction) MA-chromatography may also be a promising technique for fast biopolymer separation, which, for example, has been used to isolate a polyhistidine-tagged fusion protein (EcoR V) produced by *E. coli* bacteria [176].

Membrane adsorbers which are functionalized by the attachment of strong and weak ion-exchanger groups are another rapidly growing area of MA-chromatography. Various dead-end and cross-flow MA units as well as multi-stage isolation procedures based partially or completely on MA-chromatography have already been employed in the laboratory-scale isolation and purification of biologicals such as β-galactosidase, *Subtilisin Carlsberg*, monoclonal antibodies and antithrombin III [167, 177–183]. While true large-scale downstream processes based on MA-chromatography are still lacking, a multi-stage system suitable for processing several liters of culture supernatant per hour is discussed in Ref. 184.

5 Scale-up Considerations

In preparative and process scale chromatography the throughput, or amount of sufficiently pure product obtained per unit time and column volume, is of major importance. The development and optimization of a separation process is easier and less expensive if small columns are used, provided a reliable scaling-up method is available. The scaling up of chromatographic separation processes is a challenging problem for chromatographic engineering. Evidently a dynamic similarity must be maintained between the functions of the small and the large columns. Most experience has been gained in scaling up elution chromatography on the basis of similitude relationships [185–187]. Some data for the design of large scale columns such as: flow rate, sample load, and gradient conditions are available in the the literature [188]. For the calculations of production rates and column utilization the expressions thus derived can be consulted. The length-to-diameter ratio of the large scale columns will often be smaller and the stationary phase particles larger than in laboratory-scale liquid chromatography [189,190].

When scaling up the chromatographic process the quality of the separation ought to be maintained, which means in the simple case that the column efficiency, (i.e. the number of plates) should be about the same. This can be accomplished relatively easily when only the diameter of the column is increased and the column length and the stationary phase material remain the same. Whereas the mobile phase composition and linear flow velocity are kept constant, the volumetric flow rate of the eluent and the sample load may be increased in proportion to the cross-sectional area of the column. Keeping everything else constant, resolution and peak shape should be independent of the column [191,192]. In practice, however, a certain decrease in efficiency should be expected since it is difficult to achieve a uniform packing structure in larger scale columns [193] (Fig. 7.1). Radial temperature gradients may also be set up with a concomitant decrease in column efficiency. Columns having inner diameters greater than 5 cm should be equipped with some sort of compression device to maintain bed stability throughout the separation (Fig. 7.2). Several designs for the static or dynamic compression of the column packing have been patented [194–196]; dynamic axial compression by a piston being currently perhaps the most suitable method [194]. Dynamic compression is generally preferred to static compression, as the former allows for the adjustment of the applied pressure as the particles swell or shrink.

The choice of the particle diameter of the stationary phase will be influenced by most operating parameters when the length of the bed is changed upon scaling up or the specifications for the throughput and the desired final product quality are altered. The effect of the particle diameter on the efficiency of the column tends to be less pronounced under overloading conditions. However, an optimal ratio of column length to the square of the particle diameter can be found so that in the most common applications the optimum particle diameter is in the range from 10 µm to 30 µm [197]. Broad particle size distribution often

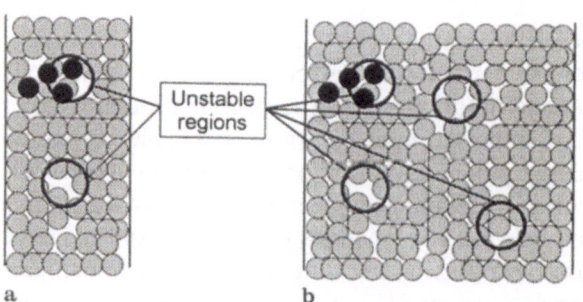

Fig. 7.1a,b. Diagrams showing the packing structure and wall support for (a) an analytical and (b) a preparative column (reproduced with permission from Ref. 194)

Unstable regions

a   b

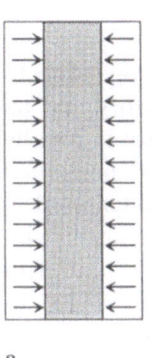 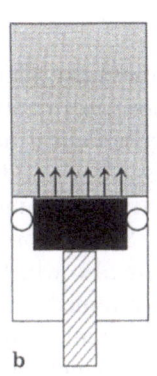 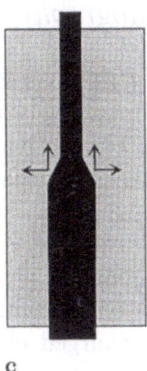

Fig. 7.2a–c. Types of compression types: (a) radial, (b) axial, (c) annular expansion (reproduced with permission from Ref. 194)

a   b   c

leads to to an inferior separation efficiency with concomitantly higher back pressure.

Production scale chromatography has been established as a powerful and versatile separation technique. Columns with inner diameters of 3.6 m and heights of 12 m have been built, mainly for separations in the oil and sugar industries [130]. Sephadex gel filtration columns with inner diameters of up to 180 cm (custom built by Pharmacia, Uppsala, Sweden) have been used to separate milk proteins, amino acids, technical enzymes and pencillins [130]. While high performance chromatography has displaced low pressure LC almost totally in the analytical field, large-scale preparative chromatography is more often than not performed at relatively low column inlet pressures. However, over the last few years production-scale high performance chromatographs have become available from a number of suppliers and the application of such instruments is on the increase in the pharmaceutical industry, especially for the purification of high-value, low-volume products, (e.g., for the separation of peptides [198]). Contrary to popular belief, high performance preparative chromatography, if properly designed and optimized, is a competitive industrial purification process. High column efficiencies and fast flow rates permit difficult separations in considerably less time, thus reducing product-degradation during separation and quite often the purification costs as well.

6 Separation Techniques in Adsorption Chromatography

Liquid chromatography is the most important purification method for biological substances when high resolution is required. Its versatility and flexibility is unsurpassed. The molecular qualities most commonly exploited in separations are size, ionic and hydrophobic properties as well as certain biospecific interactions.

6.1 Ion-Exchange Chromatography

Ion-exchange chromatography (IEC) has been the most widely applied technique in preparative protein chromatography both in the laboratory and on a production scale [199–201]. The recovery of biological activity is usually excellent. Weak anion and cation exchangers carrying diethylaminoethyl (DEAE) and carboxymethyl (CM) ligates, respectively, are most often used in both conventional and high performance IEC. Strong ion-exchangers with sulfonic acid or quarternary ammonium groups are also widely used in protein purification. In IEC, the sample components are retained by virtue of electrostatic interactions between the charged eluate molecules and the oppositely charged chromatographic surface. Consequently retention occurs on ion-exchangers when the sign of the fixed charges at the surface is the opposite of that of the net charge of the protein. However, this is only a rule of thumb, as the relationship between retention and net charge is usually not so straightforward. This is because the charge distribution over the surface of the protein molecule is not uniform and because steric effects also play an important role in determining the magnitude of the interaction. Two models currently used in protein IEC (the stochiometric displacement and the electrostatic interaction model) link the respective retention factors to the ionic strength of the mobile phase and the number of charged groups involved in the adsorption/desorption process [45, 202–207].

Gradient elution with increasing salt concentration is most widely used in the IEC of proteins. When the salt concentration of the eluent is increased, charges present on the protein molecules and at the surface of the stationary phase are screened and as a result attraction between the protein molecules and the stationary phase is diminished. Since the net charge of the protein and, in the case of the weak ion-exchangers, the charge of the chromatographic surface are both pH-dependent, control of the mobile phase pH is very important in IEC and great attention has to be paid to the nature of the buffer as well. A pH gradient may also be used for protein elution; however, due to the technical difficulties in generating smooth and reproducible pH gradients, they are less commonly employed than salt gradients.

Progress in frontal and displacement IEC has been considerable in recent years. In 1978, Torres and Peterson et al. started to develop and later optimized a system using carboxymethyldextranes [94–96, 208–210] as displacing agents. Recently, their displacer has become commercially available [211]. Other IEC protein displacers include chondroitin sulfate [87, 212], carboxymethylstarch [89], Nacolyte 7105 [213], and the polycation polyethylenimide (PEI) [93]. The latter, however, is difficult to remove from the column after the displacement run. Modified dextranes, heparin, protamin, pentosan polysulfate and block methacrylic polyampholytes have all been identified as powerful protein displacers in IEC [73, 214, 215]. Oligomeric polyvinylsulfonic acid [216] is another promising agent for protein displacement. Frontal chromatography is most widely used in on-off chromatography to purify proteins (e.g. ovalbumin and soybean trypsin inhibitor) on an anion-exchanger column [217].

6.2 Affinity Chromatography

Many biological processes involve highly specific interactions based on molecular recognition. The unique biospecificity arises from a synergistic effect of combined van der Waals, electrostatic, hydrophobic and hydrogen-bonding interactions, the effect of the aqueous medium and the complementary steric arrangement of the interacting moieties. Affinity chromatography, a term introduced by Cuatrecasas et al. in 1968 [218], exploits such biospecific interactions for separation purposes. The technique was first practiced in the form of traditional column chromatography during the 1970's. Later, it was adapted to HPLC, combining the high selectivity characteristic of biospecific interactions with the speed, efficiency, and other features of HPLC [219, 220]. It has rapidly developed into a fast, highly selective method for separating a wide variety of complex biological molecules, as well as viruses and cells. In production-scale chromatography, affinity-based separations are often seen as single-step alternatives to the multi-step processes that incorporate IEC, HIC/RPC and SEC separations [221–223].

The surface of an affinity sorbent must be highly hydrophilic without functions that would elicit nonspecific interactions [224]. The most commonly used supports are based on agarose, porous glass, silica, polyacrylamide, methacrylate and cellulose [225, 226]. Fibrous supports were developed specifically for preparative applications [227]. A biospecific ligand may be covalently linked to the stationary phase surface via a hydroxyl, amino, or carboxyl function. Frequently, a spacer arm is used to anchor the ligand to the support surface. A variety of preactivated stationary phases are commercially available to which the affinity ligands, e.g. antibodies, antigens, lectins, receptors, enzyme inhibitors, hormons or biomimetic ligands, may be attached using standard immobilization techniques. While highly specific binding is the essence of affinity chromatography, the binding between the affinant and the product to be isolated should not be too strong. Otherwise desorption and column regeneration may require harsh conditions with concomitant denaturation of the product or even the affinant itself. According to Ref. 19, triazine dyes are currently used in about 20% of all affinity chromatographic separations [228–231]. Protein A and Protein G are widely used for IgG isolation, even though they exhibit some subclass specificity [232–236]. Recent advances in genetic engineering have also helped to expand the scope of affinity chromatography. It is possible to fuse an affinity tag such as a Protein A [237] or glutathion-S-transferase [238, 239] sequence to the recombinant protein, which increases the product affinity for an immunoglobulin or glutathione column considerably and allows selective removal from most contaminants.

For product isolation, affinity chromatography is performed in the frontal mode under conditions where only the product binds to the stationary phase, while all other feed components move through the bed unretained. Nonspecific adsorption can be reduced by carefully choosing the operating conditions, i.e. the pH and salt concentration of the mobile phase and the additives used. Ligand leakage with the concomitant loss in capacity and product contamination are serious problems in affinity chromatography. In addition many affinity columns will also bind denatured or otherwise malexpressed product molecules or product fragments. Desorption is normally achieved by altering the pH, increasing the salt concentration, or introducing a chaotropic agent. Temperature changes, or reversible denaturation are also used to bring about desorption.

6.2.1 Metal Affinity Chromatography

Extending Helfferich's concept of ligand exchange to the separation of biomolecules, stationary phases with immobilized metal ions have been used in the chromatography of proteins and nucleotides and to investigate the surface topography of protein histidine residues [240–248]. The technique, called immobilized metal affinity chromatography (IMAC) or metal interaction chromatography (MIC), has even gained significance in large-scale protein purification [249]. IMAC is based on the interaction between a metal-ion electron acceptor (Lewis acid) and an electron donor (Lewis base) on the surface of a protein [241,247]. With a few exceptions, such as the interaction of phosphoproteins with Fe^{3+} ions [250,251], proteins interact through their surface histidine and – to a lesser extent – their tryptophane residues. Metal ions of intermediate polarizability such as Cu^{2+}, Ni^{2+}, Zn^{2+}, and Co^{2+} are particularly suited for interaction with proteins as they may interact not only with the nitrogen in amino- and imino-groups, but also with oxygen and sulfur [241,242]. Metal ions are immobilized on the stationary phase by chelating functions, such as the two-dentate ligand IDA (iminodiacetic acid) bond to the support. The nature of the chelating agent is of consequence. If the immobilization of the metal ion involves several coordination sites, the metal is bound strongly to the chromatographic surface and bleeding is less likely to occur. At the same time, however, the number of coordination sites available for protein binding by the stationary phase is also reduced.

Traditionally, soft agarose beads were used for protein purification by IMAC. Since then a number of rigid supports and membrane adsorbers suitable for HPLC applications including IMAC have become available [176,246, 252–256]. The mobile phase in IMAC is a buffered salt solution and the strength of the metal–protein interaction is modulated by the type and concentration of the salt [257]. Generally, retention of acidic proteins decreases, whereas the retention of basic proteins first increases, then decreases with increasing ionic strength. In preparative IMAC, the separation is achieved by differential elution with stepwise changes in the salt concentration. The pH also influences the retention behavior of proteins [249]. Elution in IMAC most commonly involves lowering the pH of the mobile phase to 6 so that the histidine residues are protonated. A mild competing agent such as glycine, histidine, or imidazole, or an organic modifier is also frequently used for the elution of the protein. Complexation of the metal ions by EDTA or the use of metal ions (which compete for the binding sites on the protein) in the mobile phase are further means for protein elution.

In IMAC, the three-dimensional protein structure is not strongly affected by chromatographic surface binding, therefore the biological activity of the product is usually well preserved. The technique has been used to purify albumin, monoclonal antibodies, immunoglobulins, blood factors, interferons, enzymes and many other proteins and polypeptides [242, 249, 258–262], usually on Zn^{2+}, Ni^{2+} or Cu^{2+} IDA columns. A displacement IMAC is described in Ref. 90. In order to facilitate the isolation of recombinant proteins by IMAC, an oligohistidine tag can be added [176, 263–266], thus imparting to them a strong affinity towards Cu^{2+}. This approach may have advantages over the use of larger tags such as Protein A, which may be difficult to remove in order to obtain the intact protein product [267, 268].

6.3 Hydrophobic Interaction Chromatography

Hydrophobic interaction chromatography (HIC) was developed in the 1970s especially for the separation of proteins using agarose-based stationary phases with a low density of mildly hydrophobic ligands [269–274]. However, most of these early stationary phases contained ionic groups as well as hydrophobic groups, thus engendering a mixed retention mechanism. More recently, rigid macroporous silica or polymeric supports have been introduced that are covered with a covalently bound hydrophilic surface layer which incorporates appropriate hydrophobic ligands, such as short alkyl, aryl or polyether chains at a comparatively low concentration [275–279]. Protein retention and selectivity depend on the nature and size of the hydrophobic moieties [280–283]. Retention is enhanced by high salt concentration in the aqueous mobile phase and therefore gradient elution with decreasing salt concentration is most commonly used in the HIC of proteins [284–286]. Due to these comparatively mild operating conditions, the molecular integrity of the native protein is normally preserved and no significant loss of biological activity occurs. For this reason, HIC is widely employed in the preparative and process-scale isolation and purification of proteins [287–294]. It should be noted, however, that some slight changes in the protein structure may occur in HIC due to a weakening of the hydrophobic forces responsible for maintaining that structure [295–297].

The energetics of the retention in HIC have been treated by adopting the solvophobic theory of Sinanoglu [298–306]. The free energy change for the transfer of the eluate molecules from the mobile to the stationary phase is evaluated by taking into account the various interactions between the components of the system. The magnitude of retention is mainly determined by the so-called cavity term, which expresses the free energy change upon adsorption; this change is due to a reduction in the molecular surface areas of both the eluate and the stationary phase that are exposed to the mobile phase. The cavity term is roughly the product of the microthermodynamic surface tension of the mobile phase and the molecular contact area upon binding. Thus retention can be considered as a solvent effect and generally the retention factor decreases with the surface tension of the mobile phase, i.e. when the salt concentration of a given eluent is lowered. Plots of protein retention factors against salt concentration, however, pass through a minimum. Retention at low salt concentrations is believed to be due to electrostatic effects [304]. At sufficiently high salt concentrations, hydrophobic interactions dominate the retention and the logarithmic retention factor increases linearly with the salt concentration; this agrees with the predictions made by the modified solvophobic theory [307]. Accordingly, the limiting slope of the log k' versus salt concentration plot is called the hydrophobic interaction parameter. Its magnitude is determined by the contact area of the protein molecule upon binding and the surface tension increment of the salt in the eluent. The molar surface tension increments of the salts follow the Hofmeister series [308]. In the case of a specific salt (magnesium- or calcium-based) binding by the proteins, significant deviations from the predictions of the simple theory based on the surface tension argument may be observed. In practice, the column temperature, the pH of the eluant and the nature of the stationary phase will also have significant effects on HIC separation [309–314].

6.4 Reversed-Phase Chromatography

Reversed-phase chromatography (RPC) was introduced in 1950 for the separation of nonpolar substances [315]. It has become the predominant branch of analytical HPLC also in the life sciences and in biotechnology [299], so that, to date, an estimated 60%–70% of all HPLC separations are carried out by RPC. Most commonly, bonded high performance stationary phases prepared by covalently binding hydrophobic ligands such as C_4-, C_8-, C_{18}-alkyl chains or aromatic functions to the surface of a rigid siliceous or polymeric support are used [316–319]. Due to the strong hydrophobic character of the stationary phase, proteins and peptides are bound very strongly from an undiluted aqueous mobile phase so that their elution requires the use of hydro-organic eluents. The separation of peptides and protein in RPC is typically carried out by gradient elution with increasing concentration of an organic modifier such as acetonitrile, methanol, tetrahydrofuran and isopropanol. In addition, the mobile phase usually contains low levels of trifluoroacetic or phosphoric acids. The role of the acids is to protonate the residual silanol groups at the surface of the siliceous support and the carboxyl groups of the eluates, as well as to form ion pairs with the charged amino groups of the substances to be separated. Ion-pairing agents such as perchlorate can be used at neutral pH.

The retention mechanisms of proteins and peptides in RPC are discussed in Refs. 320 through 326. As a rule of thumb, retention increases with size and hydrophobicity of the molecules in question. However, in the case of proteins and larger polypeptides, the three-dimensional structure, i.e., the number and location of the hydrophobic patches at the molecule surface, become predominant. It is therefore difficult to predict the retention behavior of larger peptides or proteins. Moreover, under the conditions of RPC with acidic, hydro-organic mobile phases, many proteins tend to denature and unfold either partially or completely. Oxidation, deamidation, aggregation and fractionation are also possible. The chromatographic consequences of changes in the protein structure are discussed in detail in Refs. 206 and 327 through 331.

The use of RPC in preparative work requires the refolding of the product into its native configuration after separation. This has been shown to be possible for a number of peptides and smaller proteins which are of interest to the pharmaceutical industry, such as h-insulin, h-growth hormone (hGH), tissue plasminogen activator (TPA) and various interleukins [332–334].

6.5 Hydroxyapatite Chromatography

Hydroxyapatite (HA) has been used as the stationary phase for biopolymer chromatography since 1956 [335, 336]. While these early materials were soft powders (Tiselius apatite), new production procedures were developed in the early 1980s, which resulted in HPLC compatible beads [337–340]. Recently, fluoroapatite (FA), which has similar properties to HA but significantly higher mechanical strength and chemical stability, has become available [341]. Today, HA columns are commercially available for both analytical and preparative purposes from different suppliers. HA and FA will bind both negatively and positively charged substances, yet a simple ion-exchange mechanism does not account for the observed chromatographic behavior. Both apatite compounds only bind proteins which possess an intact three-dimensional structure; therefore denaturation of a protein by chaotropic substances or by heat, largely prevents it from being adsorbed on apatite.

The physico-chemical phenomena underlying the separation were examined in the 1970's and several retention models have been proposed [342–362]. Since chromatography on FA gives similar results to chromatography on HA, it was concluded that neither the hydroxy nor the fluoro groups play a dominant role in the separation. Two types of binding sites are thus present at the chromatographic surface: the calcium ions (C-sites) and the phosphate groups (P-sites) which interact with the amino and carboxyl groups of the protein. In contact with a mobile phase of neutral pH and above, the apatite surface carries a net negative charge, as a result of a surplus of phosphate groups. This is amplified in most chromatographic separations on apatite by the use of a phosphate buffer as the mobile phase. Positively charged proteins bind by electrostatic interactions to the negatively charged surface. Desorption is brought about by a high concentration of anions, such as F^-, Cl^- and SCN^-, in the mobile phase. By a different mechanism, but with comparable efficacy, cations with a high affinity to phosphate, such as Ca^{2+} or Mg^{2+}, displace basic proteins from the surface. Negatively charged proteins are believed to bind via their carboxyl groups to the C-sites at the surface, and proteins with clusters of carboxyl groups are especially strongly bound. However, such proteins are repelled by the negatively charged apatite surface and are therefore only weakly retained under standard conditions. They are readily displaced by phosphate, fluoride or any other anion that binds strongly to calcium. Neutral proteins are eluted by specific ions such as Mg^{2+}, F^-, phosphate and, to a lesser extent, Cl^-.

The different retention mechanisms for acidic, neutral and basic proteins facilitate the development of group separation schemes according to the isoelectric points of the proteins. In processing products obtained from mammalian cell cultures, HA chromatography can be used to separate serum albumin, an acidic protein, from any more basic protein such as IgG, in a quick and simple fashion [336]. The most common protocol in protein separation on HA calls for elution in a gradient of increasing phosphate concentration. In this way, all proteins are eluted: first the acidic proteins by complexing the calcium sites and subsequently the basic proteins by charge screening [363–365]. A rather interesting alternative for downstream processing is the double gradient method, where first the basic proteins are eluted from the surface in a gradient (e.g., of increasing Cl^- concentration) followed by the elution of the acidic ones in a phosphate gradient [336, 366].

7 Applications

7.1 General

To date, chromatographic separations constitute the most important means for the isolation and purification of biotechnological products. However, a single chromatographic step will rarely suffice to obtain the pure product. The choice of which methods and sequence to select may be difficult at first, as a confusingly high number of techniques are available. Moreover, in spite of the emerging expert systems for the design [e.g. 13, 17, 337] and selection of optimal downstream procedures, the choice of methods and their sequence is at present still governed largely by personal experience and preference. Nevertheless a survey of the literature on chromatographic downstream processes has been helpful for gaining a comprehensive view on the subject. The analysis of 100 papers published in 1984 on protein purifications in *J. Biol. Chem., Biochim. Biophys. Acta, Biochem., Biochem. J., Eur. J. Biochem., Arch. Biophys., Anal. Biochem.,* and

Int. J. Biochem has been made [19]. The results of the analysis show that a typical separation scheme involves the following steps: homogenization, precipitation (in 75% of all cases by $(NH_4)_2SO_4$), ion-exchange chromatography (mostly likely on DEAE-sepharose), affinity chromatography and preparative SEC. Affinity chromatography was used in 60% of all cases. On the other hand, at that time HIC, RPC and chromatography on inorganic supports such as apatite were seldom used (Fig. 8.1).

A review published in 1992 on the purification of lipases [368] contains a similar analysis with similar results. Lipases are found in a large number of mammalian, microbial, fungal and plant species. They are rapidly gaining importance in biotechnology as flavor modifiers, industrial and diagnostic enzymes, detergent additives and digestive drugs. The analysis of 70 papers published since 1970 on microbial lipases yielded the following data (Fig. 8.2). Again, precipitation is used in 82% of all cases, usually early in purification. IEC is the most common chromatographic method (67% of all cases). Weak anion exchangers, such as those with DEAE functionalities (58%), and cation-exchangers such as with CM groups were used in 20% of all cases. Although the use of weak ion exchangers is still dominating, strong ion-exchangers are increasingly used in lipase purification. SEC was the second most popular chromatographic technique. Affinity techniques were used in 27% of the cases. Since HIC was grouped with affinity chromatography in this analysis, the significance of HIC methods cannot be quantified. In the purification of mam-

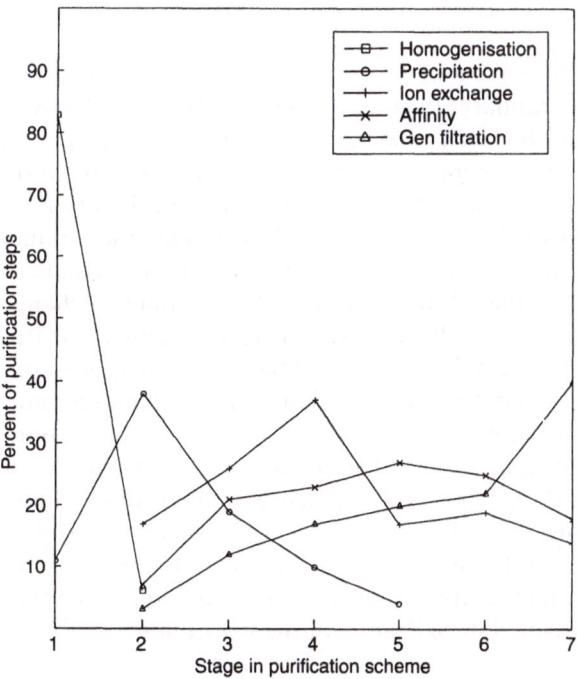

Fig. 8.1. Purification methods used at sucessive steps in biotechnological downstream processing schemes. Data for steps 8 and 9 were not included by the authors, since less than 10 % of the investigated procedures had more than 7 steps (reproduced with permission from Ref. 19)

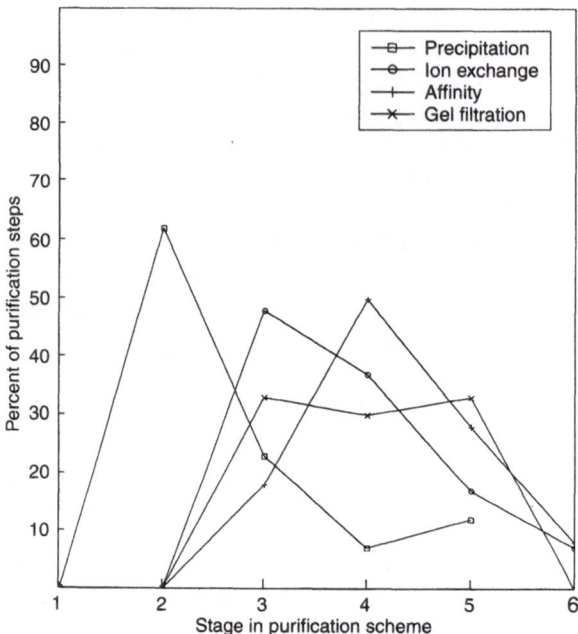

Fig. 8.2. Purification methods used at successive steps in 70 purification schemes of microbial lipases (reproduced with permission from Ref. 368)

malian lipases, which included 28 of the papers in the survey, affinity chromatography was used in over 50% of the cases. This is most likely due to the comparatively low product concentration inherent in mammalian cell cultures.

We have also examined 100 papers on protein production, published over a period of 24 months from January 1992 to January 1994, in *J. Biotech.*, *Biotech. Bioeng.*, *Appl. Microb. Biotechn.*, and *Enzym. Microb. Tech.* (Fig. 8.3). However, we found no evidence to support the assertion that new separation techniques are rapidly entering the field. In 51% of all cases, precipitation by $(NH_4)_2SO_4$ or by ethanol was used to concentrate and fractionate the product after the initial homogenization/filtration step. The use of ammonium sulfate as precipitant was favored over ethanol by a factor of six. Anion-exchange chromatography was used in 81% of all cases. DEAE-type stationary phases and strong anion-exchangers were employed to about the same extent. Approximately one-fourth of all IEC separations were cation-exchange chromatography. The relative frequency as well as the ratio of weak to strong cation-exchangers used was similar to those observed with anion-exchange chromatography. In 23% of the separations, an FPLC system was employed. Affinity chromatography was used in 21%, HIC in 12%, RPC in 13% and HA chromatography in 10% of the total number of chromatographic separations. The limited use of affinity chromatography is most likely related to the fact that over 80% of the papers investigated have dealt with the isolation and purification of microbial products.

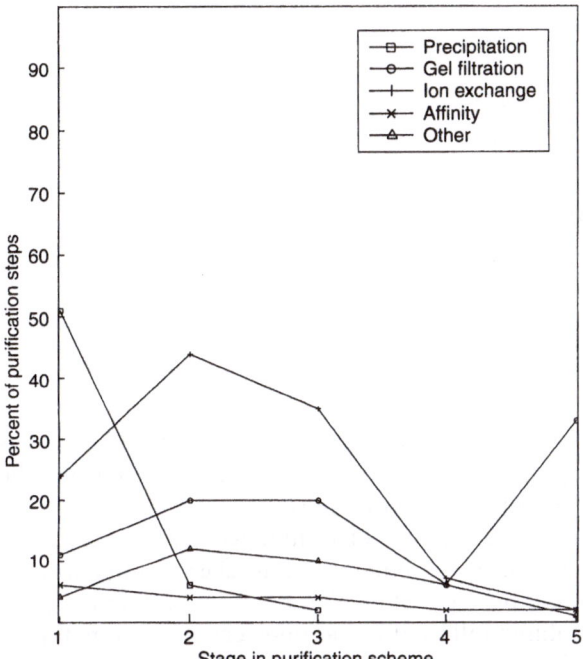

Fig. 8.3. Purification methods used at successive steps in biotechnological downstream processing schemes

7.2 Industrial Applications

High-value biotechnology products are slowly entering the market, primarily in the pharmaceutical industry. Table 1 shows a list of selected substances which are currently produced on an industrial scale. In most cases the purification schemes used in downstream processing are proprietary.

Purification strategies chosen by industry must consider ISO 9000 and other standardizations. A very efficient technique that is difficult to validate or carry out according to GMP (good manufacturing process) (e.g., one that involves an affinity column which may be difficult to sanitize or for which the validation of ligand leakage may be too involved) must sometimes be forsaken in favor of a less efficient but more reliable technique which is acceptable to the regulatory agency [369].

One of the first products for pharmaceutical use to be purified by chromatography was insulin [370, 371]. Gel filtration to remove dimers is still an important step in that purification scheme [334]. As a result of the efficiency of the downstream process, Genentech's hGH (human growth hormone) product (Protropin), produced by *E. coli*, contains less than 10 ppm of the host proteins [372]. The purification of IGF 1 (insulin-like growth factor) is described in Refs. 373 and 334. The first separation scheme takes a conventional approach. SEC is followed by IEC and then by a second SEC. A subsequent preparative isoelectric focusing step is followed by two RPC runs on C_{18}-columns to obtain the final

Table 1.

Insulin (human, bovine, porcine)
α,β,γ-Interferons
Interleukin
Human growth hormone (hGH)
Insulin-like growth factors (IGF)
Tumor necrosis factor (TNF)
Tissue plasminogen activator (t-PA)
Human serum albumin (HSA)
Blood factors (Factor VIII, Factor IX, AT III)
Monoclonal antibodies
Whey proteins
Egg white lysozym

purity required for the product. In the second scheme, recombinant IGF 1 is expressed as a fusion protein with a Protein A-tag and isolated from the supernatant by IgG-agarose affinity chromatography.

Interferons are also purified by a number of techniques depending on the source and the producing organism. Hoffmann La Roche's α-interferon is produced by *E. coli* bacteria and purified after cell separation and homogenization on an agarose affinity column followed by cation-exchange chromatography [374]. Du Pont uses Cibracron blue modified agarose to purify their β-Interferon, produced by human fibroblast cells [375]. TPA may also be obtaied from various sources. If obtained from human glioblastoma cells [376], an α-HPA-agarose column is used for isolation. If the substance is extracted from the human heart, tissue extraction and precipitation is followed by HIC, an affinity step and SEC. If human melanoma cells are the producers, IMAC, followed by a regular affinity chromatography using Concanavalin A columns and SEC give good results [377].

Merck, Sharp & Dohme used recombinant plasmids in yeast cells to produce Hepatitis B vaccine [378]. The cells are harvested, homogenized, and the surface antigen isolated on an affinity IgG-column. In Ref. 4, a four-step chromatographic method is described for the purification of recombinant hepatitis B surface antigen (r-HBsAg) from cultures of CHO-cells. The authors claim that the process is reproducible and easily adaptable for process-scale operation. After centrifuging the cells, the supernatant is passed through an HIC column, followed by a desalting SEC and an IEC. Afterwards, the product is concentrated by membrane filtration and purged of calf serum proteins by another SEC.

Monoclonal antibodies (mAb) have perhaps the broadest application spectrum of all biotechnological products. They form the basis for many immunoanalytical methods used in medicine, biochemistry, and bioanalysis in general respectively. The unique specificity of the mAb has allowed the development of in vivo therapeutics and diagnostic imaging agents. They may also serve as vehicles for targeted drug delivery. Monoclonal antibodies are produced in ascites fluid and hybridoma cell cultures. Depending on the source and the intended use, a variety of large-scale purification methods exists for these substances [2].

Currently blood plasma is the source of more than 15 therapeutic products, including albumin, immunoglobulins, various clotting factors and protease inhibitors. Over 12 million liters of plasma are processed each year in the world. Thus, plasma fractionation is perhaps the best example of how chromatographic techniques may enhance older separation methods. To date, blood is still mainly fractionated by ethanol precipitation as originally described by Cohn et al. in 1946 [379] and later modified by Kistler and Nitschmann [380]. It is unlikely that this classical process will be replaced by something different in the near future as it has many advantages. Large quantities of raw material can be processed, the involved unit operations, such as mixing and centrifugation, are straightforward, easy to scale-up, and continuously operable. The ethanol fractionation gives good yields for albumin and immunoglobulins. Moreover, ethanol is a bacteriostatic agent which also shows some potential in eliminating and deactivating viruses such as HIV and Hepatitis B. However, certain precious minor components of the blood plasma cannot be isolated by crude fractionation only. Over the years, chromatographic steps have therefore been integrated into the ethanol precipitation scheme to obtain fractions containing blood factors such as Factor III, Factor VII, Factor VIII, Factor IX, AT III and vW Factor in concentrated forms [1]. Factor VII and Factor IX may be isolated through preliminary adsorption of the cryosupernatant on DEAE-Sephadex. The effluent of this column contains the AT III which is in turn recovered by heparin affinity chromatography. The isolation of factors containing high concentrations of Fibrinogen, Fibronectin, Factor III and vW Factor is carried out by using chromatography steps with $Al(OH)_3$, DEAE-Fractogel, Heparin-Sepharose and Gelatin-Sepharose [381]. Immunoglobulins may be isolated chromatographically from either the cryosupernatant [382], from the precipitates II and III [383], or from the supernatant III [1] of the ethanol fractionation.

8 Conclusion

Advances in biotechnology depend on the availability of efficient analytical and purification methods. Chromatography has become an indispensable tool in this area. On the production scale, chromatographic separation steps are important parts of almost any protein purification scheme. On the other hand, biotechnology has provided new challenges for chromatography and contributed significantly to the development of several chromatographic techniques. Progress has been made in preserving the biological activity of the product and enhancing the efficiency of separation by the introduction of novel sorbents and an engineering approach to the design of downstream processing. As a result of favourable process economics, the affordability of many high value products of biotechnology will increase so that the general public can enjoy the benefits of biotechnology.

Acknowledgements. Some of the work presented here was supported by grant No. GM 20993 from National Institutes of Health, U.S. Public Health Service, and by grant BCS-9014119 from the National Science Foundation.

9 References

1. Burnouf T (1991) Bioseparation 1:383
2. Naveh D, Siegel RC (1991) Bioseparation 1:351
3. Nau DR (1989) BioChromatography 4:4
4. Belew M, Yafang M, Bin L, Berglöf J, Janson J-Ch (1991) Bioseparation 1:397
5. Kaul R, Mattiasson B (1992) Bioseparation 3:1
6. Ogez JR, Hodgdon JC, Beal MP, Builder SE (1989) Biotech. Adv. 7:467
7. Sadana A (1992) Bioseparation 3:145
8. Scandella C, Pettersson T (1991) Bioseparation 1:367
9. Dwyer JL (1984) Bio/Technology 2:957
10. Spalding BJ (1991) Bio/Technology 9:229
11. Lightfoot EN, Cockrem MCM (1987) Sep. Sci. Tech. 22 (2&3):165
12. Cen P, Tsao G (1993) Sep. Technol. 3:58
13. Asenjo J (1990) In: Pyle D (ed) Separations for Biotechnology 2. Elsevier London, p 519
14. Garg V, Costello M, Czuba B (1991) In: Seetharam R, Sharma S (eds) Purification and Analysis of Recombinant Proteins. Marcel Dekker, New York Basel Hong Kong, p 29
15. Ho SV (1990) In: Ladisch MR, Willson RC, Painton CC, Builder SE (Eds) Protein Purification, from Molecular Mechanism to Large Scale Processes, ACS Symposium Series 427, American Chemical Society, p 14
16. Heinrikson R, Tomasselli A (1991) In: Seetharam R, Sharma S (eds) Purification and Analysis of Recombinant Proteins. Marcel Dekker, Inc., New York, Basel Hong Kong, p 3
17. Wheelwright S (1991) Protein Purification: Design and Scale-Up of Downstream Processing. Hanser Verlag München, p 10
18. Sharma SK (1986) Sep. Sci. Tech. 21:701
19. Bonnerjea J, Oh S, Hoare M, Dunnill P (1986) Bio/Technology 4:954
20. Tswett MS (1906) Ber. Dtsch. Bot. Ges. 24:316
21. Porath J, Flodin P (1959) Nature 183:1657
22. Fischer L (1980) Gel Filtration Chromatography. In: Work TS, Burdon RH (eds) Laboratory Techniques in Biochemistry, Molecular Biology. Vol 1, Part 2, Elsevier, Amsterdam.
23. Provder T (1980) Size Exclusion Chromatography (GPC). American Chemical Society, Washington DC.
24. Andrews P (1964) Biochem. J. 91:222
25. Laurent TC, Killander J (1964) J Chromatogr. 14:317
26. Squire PG (1964) Arch. Biochem. Biophys. 107:471
27. Ackers GK (1967) J. Biol. Chem. 242:3237
28. Kovar J, Plocek J (1986) J. Chromatogr. 351:371
29. Welling GW, Vanderzee R, Welling-Wester S (1986) Trends Anal. Chem. 5:225
30. Lee AL, Velayudhan A, Horváth Cs (1989) in Durant G, Bobichon L, Florent J (eds) 8th International Biotechnology Symposium, Vol I, Societe Francaise de Microbiologie, Paris, p 593
31. Tiselius A (1943) Kolloid Z. 105:101
32. Raghavan NS, Ruthven DM (1983) AIChE J. 29:922
33. Guiochon G, Katti A (1987) Chromatographia 24:165
34. Antia FD, Horváth Cs (1989) Ber. Bunsenges. Phys. Chem. 93:961
35. Phillips MW, Subramanian G, Cramer SM (1988) J. Chromatogr. 454:1

36. Golshan-Shirazi S, Guiochon G (1992) J. Chromatogr. 603:1
37. Langmuir I (1916) J. Am. Chem. Soc. 38:2221
38. LeVan MD, Vermeulen T (1981) J. Phys. Chem. 85:3247
39. Jacobson JM, Frenz J, Horváth Cs (1987) Ind. Eng. Chem. Res. 26:43
40. Jacobson JM, Frenz J (1990) J. Chromatogr. 499:5
41. Lin B, Ma Z, Golshan-Shirazi S, Guiochon G (1989) J. Chromatogr. 475:1
42. Chen T-W, Pinto N, Brocklin L (1989) J. Chromatogr. 167
43. Huang J-X, Guiochon G (1989) J. Chromatogr. 492:431
44. Antia F (1991) Adsorption phenomena in chromatography: Studies on nonlinear behavior and the hydrophobic effect. PhD.-Thesis, Department of Chemical Engineering, Yale University, New Haven, CT, USA
45. Velayudhan A, Horváth Cs (1988) J. Chromatogr. 443:13
46. Whitley RD, Wachter R, Liu F, Wang N-HL (1989) J. Chromatogr. 465:137
47. Bellot JC, Condoret JS (1993) Proc. Biochem. 28:365
48. Bellot JC, Condoret JS (1993) J. Chromatogr. 657:305
49. Brooks CA, Cramer SM (1992) AIChE J. 38:1969
50. Frey D (1987) J. Chromatogr. 409:1
51. Van Deemter JJ, Zuyiderweg, FJ, Klinkenberg, A (1956) Chem. Eng. Sci. 5:271
52. Knox HJ, Saleem M (1972) J. Chromatogr. Sci. 10:80
53. Giddings JC (1961) J. Chromatogr. 5:61
54. Huber JFK (1969) J. Chromatogr. Sci. 7:85
55. Grushka E, Snyder LR, Knox JH (1975) J Chromatogr. Sci. 13:25
56. Horváth Cs, Lin H-J (1976) J. Chromatogr. 126:401
57. Horváth Cs, Lin H-J (1978) J. Chromatogr. 149:43
58. Kucera E (1965) J. Chromatogr. 19:237
59. Schneider P, Smith JM (1968) AIChE J. 14:762
60. Wilson EJ, Geankoplis CJ (1966) Ind. Eng. Chem. Fundamantals 5:9
61. Frey D, Schweinheim E, Horváth Cs (1993) Biotech. Prog. 9:273
62. Rodrigues AE, Lu ZP, Loureiro JM (1991) Chem. Eng. Sci. 46:2765
63. Freitag R, Frey D, Horváth Cs (1994) J. Chromatogr. 686:165
64. Tiselius A (1955) Angew. Chem. 67:245
65. Knox JH, Pyper HM (1986) J. Chromatogr. 363:1
66. Colin H (1993) In: Ganetsos G, Barker PE (eds) Preparative and Production Scale Chromatography, Chromatographic Science Series, Vol. 61, Marcel Dekker, Inc, New York – Basel – Hong Kong, p 11
67. Tondeur D, Bailly M (1993) In: Ganetsos G., Barker PE (eds) Preparative and Production Scale Chromatography, Chromatographic Science Series, Vol. 61, Marcel Dekker, Inc, New York – Basel – Hong Kong, p 79
68. Helfferich FG, Klein G (1970) Theory of Multicomponent Chromatography. Marcel Dekker, N.Y.
69. Rhee HK, Amundson NR (1982) AIChE. J. 28:423
70. Liao AW, El Rassi Z, LeMaster D, Horváth Cs (1987) Chromatographia 24:881
71. Guiochon G, Ghodbane SJ (1988) J. Phys. Chem. 92:3682
72. Viscomi G, Lande S, Horváth Cs (1988) J. Chromatogr. 440:157
73. Jayaraman G, Gadam SD, Cramer SM (1993) J. Chromatogr. 630:53
74. Cramer SM, Subramanian G (1990) Sep. Purif. Method. 19:31
75. Frenz J, Horváth Cs (1985) AIChE J. 31:400
76. Frenz J, Bourell J, Hancock WS (1990) J. Chromatogr. 512:299
77. Ramsey RS, Katti AM, Guiochon G (1990) Anal. Chem. 62:2557
78. Horváth Cs, Nahum A, Frenz J (1981) J. Chromatogr. 218:365
79. Antia FD, Horváth Cs (1991) J. Chromatogr. 556:119
80. Helfferich FG (1986) J. Chromatogr. 373:45
81. Jen SCD, Pinto N.G. (1992) J. Chromatogr. 590:3
82. Felinger A, Guiochon G (1992) J. Chromatogr. 609:35
83. Golshan-Shirazi S, Guiochon, G (1989) Anal. Chem. 61:1960
84. deBokx PK, Baarslag, PC, Urbach HP (1992) J. Chromatogr. 594:9
85. Katti A, Guiochon G (1988) J. Chromatogr. 449:25
86. Subramanian G, Cramer SM (1989) Biotech. Prog. 5:92
87. Liao AW, Horváth Cs (1988) Ann. N.Y. Acad. Sci. 589:182

88. Subramanian G, Phillips MW, Cramer SM (1988) J. Chromatogr. 439:341
89. Ghose S, Mattiasson B (1991) J. Chromatogr. 547:145
90. Kim Y, Cramer S (1991) J. Chromatogr. 549:89
91. Frenz J (1992) LC/GC Intern. 5(12):18
92. Jacobson J (1993) Recovery Processes in Biotechnology and the Role of High Performance Displacement Chromatography. Presented at the 17th International Symposium on HPLC in Hamburg Germany.
93. Frenz J, Horváth Cs (1988) In: Horváth Cs. (ed) High-Performance Liquid Chromatography: Advances and Perspectives, Academic Press, N.Y., p 211
94. Dunn BE, Edberg SE, Torres AR (1988) Anal. Biochem. 168:25
95. Torres AR, Krueger GG, Peterson E.A. (1985) Anal. Bochem. 144:469
96. Torres AR, Peterson EA (1990) J. Chromatogr. 499:47
97. Torres AR, Peterson EA (1992) J. Chromatogr. 604:39
98. Freitag R, Breier J (1994) J. Chromatogr., in press
99. Frenz J, Quan CP, Hancock WS (1991) J. Chromatogr. 557:289
100. Katti A, Dose EV, Guiochon G (1991) J. Chromatogr. 540:1
101. Bridges S, Barker PE (1993) in.: Ganetsos G, Barker PE (eds) Preparative and Production Scale Chromatography, Chromatographic Science Series, Vol. 61, Marcel Dekker, Inc, New York – Basel – Hong Kong, p 113
102. Martin, AJP (1949) Disc. Faraday Soc. 7:332
103. Fox JB, Calhoun RC, Eglinton WJ (1969) J. Chromatogr. 43:48
104. Moskvin LN, Mozzhukin AV, Tsaritsyna LG, (1975) J. Anal. Chem. USSR 30:29
105. Takahashi, Y, Goto, S (1991) J. Chem. Eng. Jpn. 24:121
106. Pronin, AY, Goryaeva, NA, Chmutov KV (1980) J. Appl. Chem. USSR 53:1125
107. Belter PA, Cunningham FL, Chen JW (1973) Biotech. Bioeng. 15:533
108. Bartels CR, Kleiman G, Korzun, Irish DB (1958) Chem. Eng. Progr. 54:49
109. Cloete FLD, Streat M (1963) Nature 200:1199
110. Janson JC, Arve (1993) Optimization of expanded bed and fixed bed adsorption of proteins. Lecture presented at the 6th European Congress in Biotechnology, Abstract W002
111. Chase HA (1994) Trends Biotech. 12:296
112. Gordon NF, Cooney ChL (1990) in Ladisch MR, Willson RC, Painton CC, Builder SE (eds) Protein Purification, from Molecular Mechanism to Large Scale Processes, ACS Symposium Series 427, American Chemical Society, p 118
113. Gordon NF, Tsujimura H, Cooney CL (1990) Bioseparation 1:9
114. Pungor E, Afeyan NB, Gordon NF, Cooney CL (1987) Bio/Technology 5:604
115. Rosensweig RE (1979) Science 204:57
116. Rosensweig RE, Lee WK, Siegel JH (1987) Sep. Sci. Technol. 22:25
117. Burns MA, Graves DJ (1985) Biotechnol. Prog. 1:95
118. Burns MA, Graves DJ (1985) Ann. NY Acad. Sci. 501:103
119. Graves DJ (1993) in: Ganetsos G, Barker PE (eds) Preparative and Production Scale Chromatography, Chromatographic Science Series, Vol. 61, Marcel Dekker, Inc, New York–Basel – Hong Kong, p 187
120. Burns MA, Graves DJ (1987) React. Polym. 6:45
121. Lochmüller CH, Wigman LS (1987) Sep. Sci. Technol. 22:2111
122. Fish BB, Carr RW, Aris R (1993) AIChE J. 39:1783
123. Ray AK, Carr RW, Aris R (1993) Chem. Eng. Sci. 49:469
124. Neuzil RW, Rosback DH, Jensen RH, Teague JR, DeRosset AJ (1980) Chem Tech August: 498
125. Ganetsos G, Barker PE (1993) In: Ganetsos G, Barker PE (eds) Preparative and Production Scale Chromatography, Chromatographic Science Series, Vol. 61, Marcel Dekker, Inc, New York – Basel – Hong Kong, p 233
126. Hashimoto K, Adachi S, Shirai Y, Horie M (1988) J. Food Eng. 8:187
127. Hashimoto K, Adachi S, Shirai Y (1988) Agric. Biol. Chem. 52:2161
128. Hashimoto K, Yamada M, Adachi S, Shirai Y (1989) J. Chem Eng. Jpn. 22:432
129. Janson J C, Jönnson J-Ä (1986), In: Janson J-C, Rydeb L (eds) Protein Purification, High Resolution Methods, Applications, VCH Publishers Inc., New York, p 35
130. Ganetsos G, Barker PE (1993) In: Ganetsos G, Barker PE (eds) Preparative and Production Scale Chromatography, Chromatographic Science Series, Vol. 61, Marcel Dekker, Inc, New York – Basel – Hong Kong, p 3
131. Freiser HH, Gooding KM (1991) J. Chromatogr. 544:125
132. Hansen SH, Helboe P, Thomsen M (1991) J. Chromatogr. 544:53

133. Chang SH, Gooding KM, Regnier FE (1976) J. Chromatogr. 125:103
134. El Rassi Z, Horváth Cs (1984) Chromatographia 19:9
135. Lloyd LL (1991) J. Chromatogr. 544:201
136. Sugii A, Harada K (1991) J. Chromatogr. 544:219
137. Yang Y-B, Regnier FE (1991) J. Chromatogr. 544:233
138. Maa Y-F, Horváth Cs (1988) J. Chromatogr. 445:71
139. Tanaka N, Hashizume K, Araki M (1987) J. Chromatogr. 400:38
140. Hjertén S, Wu B-L, Liao J-L (1987) J. Chromatogr. 396:101
141. Hjertén S, Li J-P, Liao J-L (1989) J. Chromatogr. 475:177
142. Giddings JC (1965) Dynamics of Chromatography, Part I, Principles and Theory. Marcel Dekker, New York
143. Huber J (1973) Ber. Busenges. Phys. Chem. 77:179
144. Kalghatgi K, Horváth Cs (1987) J. Chromatogr. 398:335
145. Unger KK, Jilge G, Kinkel JN, Hearn MTW (1986) J. Chromatogr. 359:61
146. Antia FD, Horváth CS (1988) J. Chromatogr. 435:1
147. Afeyan NB, Fulton SP, Regnier FE (1991) J. Chromatogr. 544:267
148. Dawkins JV, Lloyd LL, Warner FP (1986) J. Chromatogr. 352:157
149. Lloyd LL, Warner FP (1990) J. Chromatogr. 512:365
150. Afeyan NB, Gordon NF, Mazsaroff I, Varady L, Fulton SP, Yang YB, Regnier FE (1990) J. Chromatogr. 519:1
151. Regnier F (1991) Nature 350:634
152. Afeyan N, Regnier FE, Dean R (1991) US Patent 5,019,270,28 May 1991
153. Afeyan N, Fulton SP, Gordon NF, Mazsaroff I, Varady L, Regnier FE (1990) Bio/Technology 8:203
154. Fulton SP, Afeyan N, Gordon NF, Regnier FE (1991) J. Chromatogr. 547:452
155. Brandt S, Goffe RA, Kessler SB, O'Conner JL, Zale SE (1988) Bio/Technology 6:779
156. Heath CA, Belfort G (1992) Adv. Biochem. Eng. 47:45
157. Garg VK, Zale SE, Azad ARM, Holton OD (1992), In: Todd P, Sikdar SK, Bier M (eds) Frontiers in Bioprocessing II, Am. Chem. Soc. p 321
158. Piotrowski JJ, Scholla MH (1988) BioChromatography 3:161
159. Frey DD, van de Water R, Zhang B (1992) J. Chromatogr. 603:43
160. Briefs K-G, Kula M-R (1992) Chem. Engng. Sci. 47:141
161. Tennikova TB, Svec F (1993) J. Chromatogr. A 646:279
162. Svec F, Tennikova TB (1991) J. Bioactive Comp. Biopl. 6:393
163. Tennikova TB, Belenkii BG, Svec F (1990) J. Liq. Chromatogr. 13:63
164. Unarska M, Davies PA, Esnouf MP, Bellhouse BJ (1990) J. Chromatogr. 519:53
165. Tennikova TB, Bleha M, Svec F, Almazova TV, Belenkii BG (1991) J. Chromatogr. 555:97
166. Gerstner JA, Hamilton R, Cramer SM (1992) J. Chromatogr. 596:173
167. Reif O-W, Freitag R (1993) J. Chromatogr. 543:29
168. Josic D, Bal F, Schwinn H (1993) J. Chromatogr. 632:1
169. Jungbauer A, Unterluggauer F, Uhl K, Buchacher A, Steindl F, Pettauer D, Wenisch E (1988) Biotechnol. Bioeng. 32:326
170. Kikumoto Y, Hong Y-M, Nishida T, Nakai S, Masui Y, Hirai Y (1987) Biochem. Biophys. Res. Commun. 147:315
171. Nachman M, Azad ARM, Bailon P (1992) J. Chromatogra. 597:155
172. Upshall A, Kumar AA, Bailey MC, Parker MD, Favreau MA, Lewison KP, Joseph ML, Maragnora MA,, McKnight GL (1987) Bio/Technology 5:1301
173. Mandaro RM, Roy S, Hou KC (1987) Bio/Technology 5:928
174. Langlotz P, Kroner KH (1992) J. Chromatogr. 591:107
175. Krause S, Kroner KH, Deckwer W-D (1991) Biotechnol. Tech. 5:199
176. Reif O-W, Nier V, Bahr U, Freitag R (1994) J. Chromatogr. 664:13
177. Molinari R, Torres JL, Michaels AS, Kilpatrick PK, Carbonell RG (1990) Biotechnol. Bioeng. 36:572
178. Josic D, Reusch J, Köster K, Baum O, Reutter W (1992) J. Chromatogr. 590:59
179. Champluvier B, Kula M-R (1991) J. Chromatogr. 539:315
180. Josic D, Zeilinger K, Lim Y-P, Raps M, Hofman W, Reutter W (1989) J. Chromatogr. 484:327
181. Abou-Rebyeh H, Körber F, Schubert-Rehberg K, Reusch J, Josic D (1991) J. Chromatogr. 566:341
182. Lütkemeyer D, Bretschneider M, Büntemeyer H, Lehmann J (1993) J. Chromatogr. A 639:52
183. Champluvier B, Kula M-R (1993) Bioseparation 2:343

184. Reif O-W, Freitag R (1994) Bioseparation 4:369
185. Hupe KP, Lauer HH (1981) J. Chromatogr. 203:41
186. Rausch CW, Heckendorf AH High Performance Liquid Chromatography. In: Cooney CL, Humphrey AE (eds) Comprehensive Biotechnology, Vol 2, Pergamon Press, N.Y., p 537
187. Timmins RS, Mir L, Ryan JM (1969) Chem. Eng. May:170
188. Kennedy JF, White CA, Rivera Z (1988) Int. Ind. Biotechnol. 8(1):15
189. Charm SE, Matteo CC, Carlson RA (1968) Chem. Eng. Prog. Symp. Ser. 64:9
190. Janson J-C, Dunnill P (1974) In: Spencer B (ed) Industrial Aspects of Biochemistry. Part 1. North Holland Publishing Co., Amsterdam
191. Huber JFK (1975) Z. Anal. Chem. 277:341
192. Eisenbeiss F, Ehlerdi S, Wehrli A, Huber JFK (1985) Chromatographia 20:657
193. Verzele M, Dewaela C (1985) Preparative High Performance Chromatography: A Practical Guideline. RSL/Alltech Europe, Eke
194. Colin H, Hilaireau P, de Tournemire J (1990) LC/GC 8(4):40
195. Little JN, Cotter RL, Prendergast JA, McDonald PD (1976) J. Chromatogr. 126:439
196. Godbille E, Devaux P (1976) J. Chromatogr. 122:317
197. Golshan-Shirazi S, Guiochon G (1989) Anal. Chem. 61:1368
198. Scott F (1984) Process Eng., Feb: 26
199. Regnier FE (1983) In: Hearn MTW, Regnier FE, Wehr CT (eds) High Performance Liquid Chromatography of Proteins and Peptides, Academic Press, New York
200. Cramer SM, Subramanian G (1989) In: Keller EG, Yang RT (eds) New Directions in Sorption Technology. Butterworths, Stoneham, UK
201. Desplancq D, Koleman V, Chaussivert N, Rojo E, Fischer L, de Soizieu A, Egly J-M (1992) ChimicaOggi Nov/Dec:41
202. Kopaciewicz W, Regnier FE (1983) Anal. Biochem. 133:251
203. Kopaciewicz W, Rounds MA, Fasneugh J, Regnier FE (1983) J. Chromatogr. 266:3
204. Round MA, Regnier FE (1984) J. Chromatogr. 283:37
205. Drager RR, Regnier FE (1986) J. Chromatogr. 359:147
206. Regnier FE (1987) Science 238:319
207. Cysewski P, Jaulmes A, Lemque R, Sebille B, Vidal-Madjar C, Jilge G (1991) J. Chromatogr. 548:61
208. Torres AR, Peterson EA (1979) J. Biochem. Biophys. Methods 1:349
209. Peterson EA (1978) Anal. Biochem. 90:767
210. Torres AR, Edberg SC, Peterson EA (1987) J. Chromatogr. 389:177
211. Frenz J (1992) LC/GC Intern. 5(12):18
212. Lee AL, Liao AW, Horváth Cs (1988) J. Chromatogr. 443:31
213. Subramanian G, Phillips MW, Jayaraman G, Cramer SM (1989) J. Chromatogr. 484:225
214. Gerstner JA, Cramer SM (1992) Biotech. Progress 8:540
215. Gadam S, Cramer SM (1994) Chromatographia 39:409
216. Jen S-CD, Pinto NG (1990) J. Chromatogr. 519:87
217. Hodges RS, Burke TWL, Mant CT (1988) J. Chromatogr. 444:349
218. Cuatrecasas P, Wichek M, Anfinsen CB (1968) Proc. Natl. Acad. Sci. USA 61:636
219. Dean PDG, Johnson WS, Middle FA (eds) (1985) Affinity Chromatography – A Practical Approach. IRL Press, Oxford, UK.
220. Mohr P, Pommerening K (1985) In: Cazes J (ed) Affinity Chromatography: Practical and Theoretical Aspects. Marcel Dekker Inc. New York
221. Janson J-C (1982) Scaling-up of affinity chromatography, technological and economical aspects. In: Gribnau TCJ, Visser J, Nivard RJF (eds) Affinity Chromatography and Related Techniques, Elsevier, Amsterdam, p 503
222. Robinson PJ, Wheatley MA, Janson J-C, Dunnill P, Lilly MD (1974) Biotechnol. Bioeng. 16:1103
223. Labrou N, Clonis Y (1994) J. Biotechnol. 36:95
224. Morrow RM, Carbonell RG, McCoy BJ (1975) Biotech. Bioeng. 17:895
225. Yarmush ML, Colton CK (1985) Affinity Chromatography. In: Cooney CL, Humphrey AE (eds) Comprehensive Biotechnology, Vol 2, Pergamon Press, N.Y., p 507
226. Ohlsson S, Hansson L, Glad M, Mosbach K, Larsson P-O (1989) Trends Biotech. 7:179
227. Wikström P, Larsson P-O (1987) J. Chromatogr. 388:123
228. Boyer P, Hsu J (1993) Adv. Biochem. Eng. Biotech. 49:1
229. Clonis Y, Lowe C (1981) Biochim. Biophys. Acta 659:86
230. Gisch D, Reid T (1988) J. High Resol. Chromatogr. 11:258

231. Stead C (1991) Bioseparation 2:129
232. Hage DS, Walters RR, Hethcote HW (1986) Anal. Chem. 58:274
233. Hammen RF, Pang D, Remington K, Thompson H, Judd RC, Szuba J (1988) BioChromatography 3:54
234. Akerstrom B, Bjorck L (1986) J. Biol. Chem. 261:10240
235. Bjorck L, Kronvall G (1984) J. Immunol. 133:969
226. Guss B (1986) EMBO J. 5:1567
237. Sassenfeld HM (1990) Trends Biotech. 8:88
238. Smith DB, Johnson KS (1988) Gene 67:31
239. Hartman J, Daram P, Fizzell RA, Rado Th, Benos DJ, Sorscher EJ (1992) Biotech. Bioeng. 39:828
240. Helfferich FG (1961) Nature 189:1001
241. Porath J, Carlsson J, Olsson I, Belfrage G (1975) Nature 258:598
242. Porath J, Olin B, Granstrand B (1983) Arch. Biochem. Biophys. 225:543
243. Hemdan ES, Porath J (1985) J. Chromatogr. 323:255
244. Lönnerdal B, Keen CL (1982) J. Appl. Biochem. 4:203
245. Fanou-Ayi L, Vijayalakshmi M (1983) Ann. NY. Acad. Sci. 413:300
246. El Rassi Z, Horváth Cs (1986) J. Chromatogr. 359:241
247. Sulkowski E (1985) Trends Biotechnol. 3:1
248. Hemdan ES, Zhao Y-J, Sulkowski E, Porath J (1989) Proc. Natl. Acad. Sci. USA 86:1811
249. El Rassi Z, Horváth Cs (1990) In: Gooding KM, Regnier FE (eds) HPLC of Biological Macromolecules Methods and Applications. Marcel Dekker Inc., N.Y. p 179
250. Anderson L, Porath J (1986) Anal. Biochem. 154:250
251. Ramadan N, Porath J (1985) J. Chromatogr. 321:93
252. Seshadi T, Kampschulze U, Kettrup A (1980) Z. Anal. Chem. 300:124
253. Gimpel M, Unger K (1983) Chromatographia 17:200
254. Takayanagi H, Hatano D, Fujimura K, Ando T (1985) Anal. Chem. 57:1840
255. Corradini D, El Rassi Z, Horváth Cs, Guerra G, Horn W (1988) J. Chromatogr. 458:1
256. Bonn G, Kalgathgi K, Horne W, Horváth Cs (1990) Chromatographia 30:484
257. Porath J (1987) Biotechnol. Progr. 3:14
258. Maisano F, Testori SA, Grandi G (1989) J. Chromatogr. 472:422
259. Andersson L, Sulkowski E, Porath J (1987) Cancer Res. 47:3624
260. Andersson L (1984) J. Chromatogr. 315:167
261. Andersson L, Sulkowski E, Porath J (1991) Bioseparation 2:15
262. Kastner M, Neubert J (1991) J. Chromatogr. 587:43
263. Gentz R, Chen C-H, Rosen CA (1989) Proc. Natl. Acad. Sci. USA 86:821
264. Moks T, Abrahmsen L, Österlöf B, Josephson S, Oestling M, Enfors SO, Persson I, Nilsson B, Uhlen M (1987) Bio/Technology 5:379
265. Germio J, Bastia D (1984) Proc. Natl. Acad. Sci. USA 81:4692
266. Sassenfeld HM, Brewer SJ (1984) Bio/Technology 2:76
267. Hochuli E, Bannwarth W, Döbeli H, Gentz R, Stüber D (1988) Bio/Technology 6:1321
268. Janknecht R, de Martynoff G, Lou J, Hipskind RA, Nordheim A, Stunnenberg HG (1991) Proc. Natl. Acad. Sci. USA 88:8972
269. Yon RJ (1972) Biochem. J. 126:765
270. Shaltiel S, Er-el Z (1973) Proc. Natl. Acad. Sci. USA 70:778
271. Hjertén S, Rosengren J, Pahlman S (1974) J. Chromatogr. 101:281
272. Hofstee BJH (1973) Anal. Biochem. 52:430
273. Hjertén S (1973) J. Chromatogr. 87:325
274. Hjertén S, Yao K, Liu ZQ, Yang D, Wu BL, Liao J-L (1986) J. Chromatogr. 354:203
275. Engelhard H, Schön U (1986) J. Liq. Chromatogr. 9:3225
276. El Rassi Z, Horváth Cs. (1986) J. Liq. Chromatogr. 9:3245
277. Janzen R, Unger K, Giesche H, Kinkel JN, Hearn MTW (1987) J. Chromatogr. 397:91
278. Kato Y, Kiramura T, Hashimoto T (1984) J. Chromatogr. 298:407
279. Keda T, Yasui Y, Ishida Y (1987) Chromatographia 24:427
280. Shaltiel S (1974) Hydrophobic Chromatography. In: Jakoby WB, Wilchek M (eds) Methods in Enzymology, Vol. 34, Academic Press, N.Y., p 126
281. Alpert AJ (1986) J. Chromatogr. 359:85
282. Hofstee BJH (1976) Hydrophobic Adsorption Chromatography of Proteins. In: Catsinpoolas N (ed) Methods of Protein Separation, Vol. 2, Plenum Press, N.Y., p 245
283. Gooding DL, Schmuck MN, Gooding KM (1984) J. Chromatogr. 296:107

284. Fausnaugh I, Regnier FR (198) J. Chromatogr. 359:131
285. Floyd T, Hartwick R (1986) In: Horvath Cs (ed) High Performance Liquid Chromatography: Advances and Perspectives. Academic Press New York, Vol 4, p 45
286. Hjerten S (1976) In: Catsinpoulas N (ed) Methods of Protein Separation, Vol. 2, Plenum Press New York, p 233
287. Miller NT, Feibush B, Karger BL (1985) J. Chromatogr. 316:519
288. Goheen SC, Matson RS (1985) J. Chromatogr. 326:235
289. Josic D, Schütt W, Renswoude JV, Reuter W (1986) J. Chromatogr. 353:13
290. Corradini D, Giardi MT, Massacci A (1987) J. Chromatogr. 395:523
291. Berkowitz SA, Henry MP (1987) J. Chromatogr. 389:317
292. Dunhill P (1983) Process Biochem. 18(5):9
293. Goheen SC, Engelhorn SC (1984) J. Chromatogr. 317:55
294. Chang CT, McCoy BJ, Carbonell RG (1980) Biotech. Bioeng. 12:377
295. Ingraham RH, Lau SYM, Taneja AK, Hodges RS (1985) J. Chromatogr. 327:77
296. Wu SL, Benedek K, Karger BL (1986) J. Chromatogr. 359:3
297. Wu SL, Figueroa A, Karger BL (1986) J. Chromatogr. 371:3
298. Sinanoglu O (1967) Intermolecular Forces in Liquids. In: Hirschfelder JO (ed) Advances in Chemical Physics, Vol. 12, Wiley, N.Y., p 283
299. Melander WR, Horváth Cs (1980) Reversed Phase Chromatography. In: Horváth Cs (ed) High Performance Liquid Chromatography – Advances and Perspectives, Vol. 2, Academic Press, N.Y., p 114
300. Melander WR, Horváth Cs (1980) Thermodynamics of Hydrophobic Adsorption. In: Suffet IH, McGuire MJ (eds) Activated Carbon Adsorption of Organics from the Aqueous Phase, Vol.1, Ann Arbor Science, MI, p 65
301. Sinanoglu O, Fernandez A (1985) Biophys. Chem. 21:167
302. Sinanoglu O (1981) J. Chem. Phys. 75:463
303. Heron S, Tchapla A (1991) J. Chromatogr. 556:219
304. Melander WR, El Rassi Z, Horváth Cs (1989) J. Chromatogr. 469:3
305. Melander WR, Horváth Cs (1977) Arch. Biochem. Biophys. 183:200
306. Melander WR, Horváth Cs (1977) J. Sol. Phas. Biochem. 2(2):141
307. Melander WR, Corradini D, Horváth Cs (1984) J. Chromatogr. 317:67
308. Hofmeister F (1888) Arch, Exp. Pathol. Pharmakol. 24:247
309. Arakawa T, Timasheff SN (1984) Biochemistry 23:5912
310. Barford RA, Kumosinski TR, Parris N, White AE (1988) 458:57
311. Kitamura T, Kato Y (1985) J. High Resol. Chromatogr. Chromatogr. Commun. 8:306
312. Goward G, Atkinson T, Scawen MD (1986) J. Chromatogr. 369:235
313. Schmuck MN, Nowlan MP, Gooding KM (1986) J. Chromatogr. 371:55
314. Wetlaufer DB, Koenigbauer MR (1986) J. Chromatogr. 359:55
315. Howard GS, Martin AJP (1950) Biochem. J. 46:532
316. Unger K, Lork KD (1988) Eur. Chromatogr. News 2(2):14
317. Unger K, Ansprach B, Janzen R, Jilge G, Lord KD. In: Horváth Cs (ed) High Performance Liquid Chromatography, Advances and Perspectives, Vol 5, Academic Press, San Diego, p 2
318. Freiser HH, Gooding KM (1987) BioChromatography 2(4):186
319. Stuurman ▌, Köhler J, Janson J, Litzen A (1987) Chromatographia 23:341
320. Mann D, Moreno R (1987) Prep. Biochem. 14:91
321. Barford R, Sliwinski B, Breyer A, Rothbart H (1982) J. Chromatogr. 235:281
322. Frenz J, Hancock W, Henzel W, Horváth Cs (1990) In: Gooding K, Regnier F (eds) HPLC of Biological Macromolecules, Marcel Dekker, New York, p 145
323. Melander WR, Chen B-K, Horváth Cs (1979) J. Chromatogr. 185:99
324. Carr P, Li J, Dallas AJ, Eilkens DI, ChooTan L (1993) J. Chromatogr. 656:113
325. Snyder LR, Stadalius MA (1986) In: Horváth Cs (ed) High Performance Liquid Chromatography, Advances and Perspectives, Vol 4, Academic Press, San Diego, p 195
326. Geng X, Regnier F (1984) J. Chromatogr. 296:15
327. Kin S, Karger BL (1990) J. Chromatogr. 499:89
328. Lu XM, Figuerosa A, Karger BL (1988) J. Am. Chem. Soc. 110:1978
329. Cohen SA, Benedek K, Tapuhi Y, Ford JC, Karger BL (1985) Anal. Biochem. 144:275
330. Benedek K, Dong S, Karger BL (1984) J. Chromatogr. 317:227
331. Benedek K, Karger BL (1986) J. Chromatogr. 359:19
332. Champney S (1990) J. Chromatogr. 522:163
333. Seipke G, Mullner H, Grau U (1986) Angew. Chem. Int. ed Engl. 25:535

334. Sofer GK (1986) Biotechnology 8:712
335. Tiselius A, Hjertén S, Levin Ö (1956) Arch. Biochem. Biophys. 65:132
336. Hjerten S, Lindeberg J, Shopova B (1988) J. Chromatogr. 440:305
337. Kadoya T, Isobe T, Ebihara M, Ogawa T, Sumita M, Kuwahara H, Kobayashi A, Ishikawa T, Okuyama T (1986) J. Liq. Chromatogr. 9:3542
338. Kasai R, Yamaguchi H, Tanaka O (1987) J. Chromatogr. 407:205
339. Kawasaki T (1990) J. Chromatogr. 544:147
340. Kawasaki T, Takahashi S, Ikeda K (1985) Eur. J. Biochem. 152:361
341. Sato T, Okuyama T, Ogawa T, Ebihara M (1989) Bunseki Kagaku 38:34
342. Bernardi G, Kawasaki T (1968) Biochim. Biophys. Acta 160:301
343. Bernardi G (1971) Methods Enzymol. 21:95
344. Bernardi G (1971) Methods Enzymol. 22:325
345. Bernardi G (1973) Methods Enzymol. 27:471
346. Gorbunoff MJ (1984) Anal. Biochem. 136:425
347. Gorbunoff MJ (1984) Anal. Biochem. 136:433
348. Gorbunoff MJ, Timasheff SN (1984) Anal. Biochem
349. Kawasaki T, Takahashi S, Ikeda K (1985) Eur. J. Biochem. 152:361
350. Kawasaki T, Ikeda K, Takahashi S, Kuboki Y (1986) Eur. J. Biochem. 155:249
351. Kawasaki T (1988) Sep. Sci.Technol. 23:1105
352. Kawasaki T, Niikura M, Kobayashi Y (1990) J. Chromatogr. 515:91
353. Kawasaki T, Niikura M, Kobayashi Y (1990) J. Chromatogr. 515:125
354. Inuoue S, Ohtaki N (1990) J. Chromatogr. 515:193
355. Kawasaki T (1991) J. Chromatogr. 544:147
356. Bernardi G, Giro M, Gaillard C (1972) Biochim. Biophys. Acat 278:409
357. Gorbunoff M (1980) J. Chromatogr. 187:224
358. Gorbunoff M (1985) Meth. Enzymol. 11:370
359. Kawasaki T (1978) J. Chromatogr. 151:95
360. Kawasaki T (1978) J. Chromatogr. 157:7
361. Kawasaki T (1988) Sep. Sci. Technol. 23:617
362. Kawasaki T (1988) Sep. Sci. Technol. 23:1105
363. Kikkawa U, Ono Y, Ogita K, Fujii T, Asaoka Y, Sekiguchi K, Kosaka Y, Igarashi K, Nishizuka Y (1987) FEBS Lett. 217:227
364. Hatayama T, Fujio N, Yukioka M, Funae Y, Kinoshita H (1989) J. Chromatogr. 481:403
365. Yamakawa Y, Miyasaka K, Ishikawa T, Yamada Y, Okuyama T (1990) J. Chromatogr. 506:319
366. Kawasaki T, Kobayashu W, Ikeda K, Takahashi S, Monma H (1986) Eur. J. Biochem. 157:291
367. Leser E, Asenjo J (1994) In: Pyle D (ed) Separations for Biotechnology 3, SCI Roy. Soc. Chem., Cambridge, UK, p 260
368. Taipa MA, Aires-Barros MR, Cabral JMS (1992) J. Biotech. 26:111
369. Wisniewski R (1992) Bioseparation 3:77
370. Kroeff E, Owens R, Campbell E, Johnson R, Marks H (1989) J. Chromatogr. 461:45
371. McLeod A, Auf Der Mauer A, Wood S (1990) J. Chromatogr. 502:325
372. Hancock W (1990) In: Hancock W (ed) HPLC in Biotechnology Wiley Interscience, New York, p 1
373. Cornell HJ, Boughdady NM, Herington AC (1984) Prep. Biochem. 14:123
374. Staehelin T, Hobbs DS, Kung H, Lai C-Y, Pestka S (1981) J. Biol. Chem. 256:9750
375. Knight E, Fahey D (1981) J. Biol. Chem. 256:3609
376. Nielsen LS, Hanson JG, Skriver L, Wilson EL, Kaltoft K, Zeuthen J, Dano K (1982) Biochemistry 21:6410
377. Dingeman RC, Collen D (1981) J. Biol. Chem. 256:7035
378. McAleer WJ, Buynak EB, Maigetter RZ, Wampler DE, Miller WJ, Hilleman MR (1984) Nature 307:178
379. Cohn EJ, Strong LE, Hughes WL, Mulford DJ, Ashworth JN, Melin M, Taylor HL (1946) J. Am. Chem. Soc. 8:159
380. Kistler P, Nitschmann HS (1962) Vox Sang. 7:414
381. Burnouf-Radosevich M, Burnouf T (1988) French Patent Application No 8807530
382. Saint-Blankart J, Kirzin JM, Riberon P, Oetit F, Fourcart J, Girot P, Boschetti E (1982) In: Gribnau TCJ, Visser J, Nivard JF (eds) Affinity Chromatography and Related Techniques. Elsevier Scientific Publishing Company, Amsterdam, p 305
383. Tousch D, Allary M, Saint-Blancard J, Boschetti E (1989) In: Stoltz JF, Rivat C (eds) Biotechnology of Plasma Proteins, 175:229, Colloque Inserm.

Extractive Bioconversion of Lactic Acid

Pradip K. Roychoudhury, Aradhana Srivastava and Vikram Sahai
Department of Biochemical Engineering & Biotechnology,
Indian Institute of Technology, New Delhi - 110016, India

Lactic acid production has received much attention due to its numerous applications in the food and biochemical industries. The major industrial carbon substrates generally used in the bioconversion are molasses, whey permeate, corn, potato, sulphite waste liquor, etc. End-product inhibition in lactic acid bioconversion causes several problems. Amongst them, the most important ones are the low lactate formation rate and its recovery from the bioreaction mixture. To solve these problems, an integrated approach of bioconversion with separation has received considerable attention using different extraction techniques (ion-exchange, liquid-liquid extraction, membranes) for the separation. The significant role of process parameters to enhance the lactic acid productivities in extractive bioconversion processes are discussed in this review.

Advances in Biochemical Engineering/
Biotechnology, Vol. 53
Managing Editor: A. Fiechter
© Springer-Verlag Berlin Heidelberg 1995

List of Symbols and Abbreviations

A_m	= membrane area available for dialysis (m^2)
b	= volumetric flow rate of base (1 min^{-1})
D_R	= dilution rate in bioreactor (h^{-1})
D_S	= dilution rate in extractor (h^{-1})
f	= volumetric flow rate of nutrient (1 min^{-1})
F_D	= flow rate into and out of dialysis (1 min^{-1})
F_F^0	= flow rate into the bioreactor circuit (1 min^{-1})
F_R	= medium recycle flow rate (ml min^{-1})
F_S	= solvent flow (ml min^{-1})
K_d	= mortality constant (h^{-1})
K_p	= product inhibition constant
K_s	= substrate limitation constant (g l^{-1})
n	= cell concentration (g l^{-1})
$N(t)$	= specific growth rate (h^{-1})
$P(t)$	= specific lactic acid production rate (h^{-1})
P_0, P_1	= lactic acid concentration at the exit and inflow stream (g l^{-1})
P_d	= product concentration in the dialysate circuit (g l^{-1})
P_F^0	= product concentration in the bioreactor feed (g l^{-1})
pH, pH$_e$	= pH in the bioreactor and extractor respectively
P_{mp}	= permeability of membrane to product
P_{ms}	= permeability of membrane to substrate
P_s	= lactic acid concentration in solvent (g l^{-1})
$Q_p X$	= volumetric productivity based on total volume (gel+mesh)
S_d	= substrate concentration in the dialysate circuit (g l^{-1})
S_f	= substrate concentration in the bioreactor circuit (g l^{-1})
S_f^0	= substrate concentration in the bioreactor feed (g l^{-1})
t	= time (h)
V_0	= bioreaction mixture volume (1)
V_1	= tank D volume (1)
V_d	= liquid volume in dialysate circuit (1)
V_f	= liquid volume in bioreactor circuit (1)
V_F	= bioreactor volume (1)
V_R	= volume of reservoir (1)
X_f	= cell-mass concentration in bioreactor circuit (g l^{-1})
$Y_{p/s}$	= g of lactic acid produced per g of lactose consumed
Y_x, Y_m	= input mathematical parameters of the linear system
$Y_{x/s}$	= g of cells produced per g of lactose consumed
Z_e	= nitrogen concentration in the influent (g l^{-1})
α, β	= Luedeking and Piret model constant
α_1	= substrate/cell ratio
β_1	= specific maintenance rate (h^{-1})
Γ	= product/substrate ratio
μ_m	= maximum specific growth rate (h^{-1})

1 Introduction

Lactic acid was first discovered as a bioconversion product by Blondeau in 1847 [1]. It was investigated by Pasteur as one of his first microbiological problems. Lactic acid occurs in two optically active forms, the D(−) and the L(+) isomers, and one optically inactive DL form. Lactic acid is usually present in the DL form (a racemic mixture) during growth. Normally, pyruvic acid, formed as an intermediate, is acted on by either a D-lactate dehydrogenase or a L-lactate dehydrogenase to produce D(−) lactic acid or L(+) lactic acid. The physical and chemical properties of lactic acid production have been reviewed by Vickroy [2].

The most serious problem of the lactic acid process is end-product inhibition. To circumvent this problem, an integrated approach with separation of the products has received considerable attention. Various separation techniques have been employed [3–25]. Separation by solvent extraction is not preferred because it causes several physical, chemical and biochemical effects on the catalytic activity of microbial cells [7]. The membrane separation technique seems to be unimpressive because membrane fouling by non-diffusible and other components of the bioreaction mixture poses operational problems. In addition, membranes are expensive and their lifetime is hard to predict [8]. In both the separation techniques, the concentration of lactic acid in the bioreaction mixture is generally maintained at much lower values, which are below the concentration required for complete inhibition ($> 80 \text{ g l}^{-1}$) to minimize the effects of end-product inhibition on cell growth by lactic acid. This constraint, however, results in increased recovery costs arising from a low lactate concentration in the process mixture.

A novel concept for circumventing end-product inhibition involves the use of an ion-exchange resin in direct contact with the bioreaction mixture [3]. This technique permits both cell growth and lactic acid biosynthesis even in presence of very high lactic acid concentrations ($80–265 \text{ g l}^{-1}$) in the bioreaction mixture by providing inhibition-free spaces around the resin particles. This is a desirable feature from the point of view of lactic acid production and requires a detailed investigation before it can be applied in industrial processes. Therefore, studies were planned using the ion-exchange resin in the lactic acid extractive process.

In the extractive process, substrate and cells in the culture medium can be recycled repeatedly. Because of the reduced restrictions imposed by lactic acid, the conversion of the substrate to lactic acid would increase tremendously. In this process, the productivity can be increased by increasing the throughput of substrate.

2 Lactic Acid Biosynthesis

Lactic acid bioconversion follows the EMP pathway [26–28]. The net reaction in the transformation of glucose to lactic acid is given as follows:

$$\text{glucose} + 2\ P_i + 2\ ADP \rightarrow 2\ \text{lactate} + 2\ ATP + H_2O$$

Lactate is formed from pyruvate by a variety of microorganisms. The reduction of pyruvate by NADH to form lactate is catalyzed by lactate dehydrogenase. The regeneration of NAD^+ in the reduction of pyruvate to lactate sustains the continued operation of glycolysis under anaerobic conditions. If NAD^+ were not regenerated, glycolysis would not proceed beyond glyceraldehyde 3-phosphate which means that no ATP will be generated. Consequently, after the exhaustion of the ATP pool, even the first intermediate of the glycolysis, i.e. glucose-6-phosphate cannot be formed and the whole process will stop. The hexokinase, phosphofructokinase and the pyruvate kinase participate in regulating the pace of glycolysis. In the metabolic pathway, enzymes catalysing the essentially irreversible reactions are potential sites of control. Phosphofructokinase is the most important control element for the glycolytic pathway. Phosphofructokinase is inhibited by H^+ ions, which ultimately prevents excessive formation of lactate [27].

A wide range of microorganisms have been used for lactic acid biosynthesis. The selection of the microorganisms depends upon the type of substrate used for the process. Commercially, cane and beet sugar, whey and hydrolysed starch are used for lactic acid biosynthesis. Dextrose from corn starch was the most commonly used feedstock in the late 1950s [29]. Concentrated whey is also employed without pretreatment by the Sheffield Product Co. of Norwich, NY, USA. Potato starch was employed in Germany for lactic acid production on an industrial scale [29]. Prescott and Dunn [30] have summarized the bacteria which produce lactic acid in large quantities. These bacteria may be classified as homofermentative (producing lactic acid and cells) and heterofermentative (producing lactic acid, cells and some other products such as ethanol, carbon dioxide, acetic acid and glycerol). Only the homofermentative organisms are of industrial importance for lactic acid manufacture for obvious reasons. The homofermentative bacteria are from the genera *Lactobacillus*, *Streptococcus* and *Pediococcus* [31]. These organisms grow optimally at a temperature of about 40°C and pH in the range 5–7. These are facultative anaerobes but do not use respiration for ATP generation. The selection of the microorganisms depends primarily upon the carbohydrate to be bioconverted. Gasser [32] has tabulated different *Lactobacilli* growing on different sugars (Table 1). *Lactobacillus bulgaricus*, *Lactobacillus casei* and *Streptococcus lactis* are used for bioconversion of lactose [26]. Adapted strains of *Lactobacillus delbrueckii* and *Lactobacillus leichmanni* are typically employed for the bioconversion of glucose [26]. *Lactobacillus pentosus* has been used for the bioconversion of sulphite waste liquor [33]. Nakamura and Crowell [34] have isolated the homofermentative strain called *Lactobacillus amylophilus*

Table 1. *Lactobacillus* spp grown on different substrates

Substrates	Microorganisms	References
Glucose	*L. leichmanni* *L. delbrueckii* *S. cremoris*	Vickroy [2]; Abhorey and Williamson [38]
Lactose	*L. bulgaricus* *L. casei* *S. lactis*	Vickroy [2]
Molasses or sucrose	*L. delbrueckii*	Srivastava et al. [3]; Aksu and Kutsal [2]
Sulphite waste liquor	*L. pentosus*	Leonard et al. [33]
Starch	*L. amylophilus*	Nakamura and Crowell [34]

which is capable of bioconverting starch to L(+) lactic acid with 90% yields. Pure cultures, as well as a mixture of strains, have been used for commercial production [35–38]. The fungus *Rhizopus orhyzae* produced L(+) lactic acid and could also utilize starch feedstock [26].

Lactic acid bacteria are extremely limited in their synthetic capabilities. They always require a large number of growth factors such as amino acids and vitamins for growth. Koser [37] has reviewed the nutritional requirement of *Lactobacilli*.

3 Problems Associated with Lactic Acid Bioconversion

The basic problem in lactic acid bioconversion is end-product inhibition which in turn causes several problems in the bioconversion process. Among these, the most important ones are, (i) low cell concentration, (ii) low lactate formation rate and its recovery from the bioreaction mixture. These factors result in very low productivities with a typical value of $0.016 \ g \, l^{-1} h^{-1}$ in conventional batch process [3] and make the process economically unattractive. Another problem in lactic acid bioconversion is due to highly corrosive nature of lactic acid. This poses the additional requirement of corrosion-resistant materials for bioreactors.

Recovery processes dealing with bioreaction mixtures containing high concentrations of lactate (more than 20% w/v) require special attention because such solutions are unstable: the lactic acid is easily converted to its polymeric form and produces a cyclic dimer (lactide) or linear polymers of various chain lengths [2]. Also lactic acid may undergo self-esterification because of the hydroxyl and carboxylic groups present in its molecule. Like polymerization, self-esterification is also more pronounced at high lactate concentrations.

Amongst the homofermentative organisms, it is difficult to make a carbon balance in the case of *L. delbrueckii* grown in a batch process. The

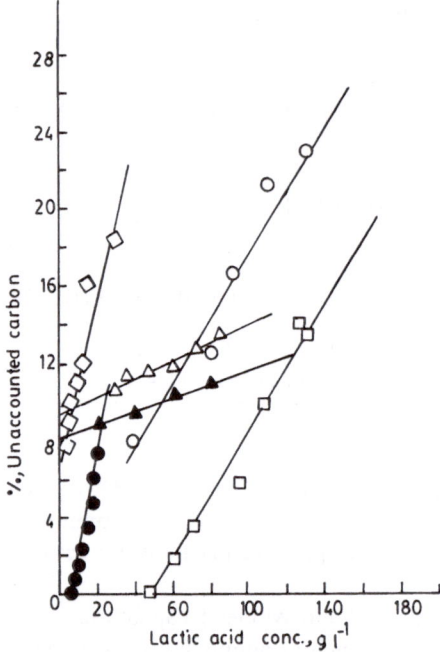

Fig. 1. Relationship between unaccounted carbon and lactic acid concentration in the bioreaction mixture. Data sources: ● Buyukgungor et al., ○ Gonclaves et al., □ Mahaia and Cheryan, ◇ Rogers et al., △ Ishizaki and Ohta, ▲ Srivastava et al.

carbon consumed should appear in the cells and the lactic acid. However, it has been possible to account for the carbon consumed only to the extent of about 85–90%. The carbon which remains unaccounted for, increases in proportion to the concentration of lactate in the reaction mixture (Fig. 1). The carbon, present in proteins which appear as enzymes in the bioreaction mixture, would account for only 0.2% of the total carbon consumed. On the other hand, if the remaining carbon were to appear in cells, it would mean that the cells should have up to 61.9% carbon. This value is, however, much higher than its expected normal value for bacteria. Efforts to identify the presence of components other than lactic acid in culture reaction mixtures of this organism have not been successful.

4 Extractive Bioconversion

It is generally agreed that the growth and activity of microorganisms in a batch culture may be limited either by the availability of nutrients or by the accumulation of toxic materials. These toxic materials which act as growth inhibitors are normally metabolic products. Thus, if one can remove the toxic metabolites from the reaction mixture, one may expect to achieve a higher cell population and to project growth and activity patterns into areas not usually attainable. In order to remove the effects of inhibitor accumulation, some methods for separating off the inhibitor during cultivation would have to be employed. Such methods would have to be selective for the inhibitor and capable of being operated aseptically. The inhibitor may be made to leave the system physically or may be kept trapped within the system by a suitable separation technique. Extractive bioconversion incorporates separation of the inhibitor product during bioconversion and thus improves the process productivities substantially. The separation technique may involve use of any one of the following:

(a) adsorption,
(b) ion exchange,
(c) liquid-liquid extraction,
(d) aqueous two-phase system,
(e) membrane separation.

The efficiency of a particular extractive bioconversion process depends upon the separation technique adopted. The material used for the extraction should be nontoxic, biocompatible, thermostable, reusable, and should have a high loading capacity and selectivity for the desired product.

4.1 Adsorption

The general arrangements for separation by adsorption are shown in Fig. 2. The adsorbent saturated with lactic acid leaves the system continuously and thus keeps the inhibitor concentration low in the reaction mixture. Porous solid adsorbents

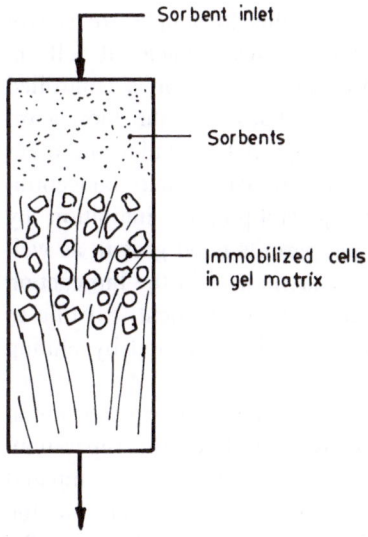

Sorbent inlet

Sorbents

Immobilized cells
in gel matrix

Sorbent outlet

Fig. 2. General arrangement of the extractive bioconversion process coupled with adsorption [6, 42, 43]

providing a large surface area are used for in-situ product separations. They effectively remove end-product inhibition and enhance process productivity [4, 5, 39–46]. Generally, activated carbon [39, 40] and polymeric resins [4, 5, 41, 42, 44] are employed as adsorbents. Activated carbon coated with a biofilm of the homo-lactic organism, *Streptococcus thermophilus*, in a fluidized bed reactor, effectively extracted lactic acid from a bioreaction mixture, resulting in an increased productivity of $12 \, \mathrm{g} \, l^{-1} h^{-1}$ under uncontrolled pH conditions [43]. Davison et al. [6] used a fluidized bed bioreactor containing immobilized organisms. A stream of denser adsorbent particles that fall through the bed, adsorb the inhibitory products including lactic acid from the bioreaction mixture and are removed from the base of the column bioreactor. The pH was controlled by sorbent addition, eliminating the need for conventional types of pH controllers. The authors reported that the inhibition of the bacteria by lactate was moderate in their system, resulting in increased glucose conversion.

The main disadvantage of the separation using adsorbent is its toxicity to microbes. Besides this, the limited capacity of adsorbent, when whole bioreaction mixture is contacted, can make this process unattractive because of the following factors: (a) non-specific adsorption of the bioreaction mixture components, (b) attrition of the adsorbent because of irreversible adsorption, and (c) the sugar adsorbed becomes caramelized when heat is used to regenerate the adsorbent. These are inherent problems with adsorbents and cannot be easily resolved.

4.2 Ion-Exchange

Ion exchangers have been employed for the recovery of various products such as carboxylic acids, volatile acids and antibiotics in extractive bioconversions [3, 47–49]. Recently ion-exchange resin has also been used [3] in extractive lactic acid bioconversion. The general arrangements for this technique are shown in Fig. 3. A bioreaction mixture, containing actively growing bacterial cells in a bioreactor, is allowed to pass through a resin-packed column in a controlled manner and the effluents of the column are recycled back into the bioreactor. The cells thus remain in direct contact with the resin particles. Unlike the other extractive processes techniques which maintain a low lactic acid concentration in the bioreaction mixture for circumventing end-product inhibition by removing lactic acid from the system, this technique permits good growth as well as the accumulation of a very high concentration of lactic acid in the bioreaction mixture, mainly due to the formation of a less inhibited environment, existing in a liquid film around the resin particles. The lactic acid is recovered by eluting the resin column at the end of the process.

The lactic acid yield by *L. delbrueckii* was enhanced when an anion exchange resin (Amberlite IRA 400, OH⁻-form) was used for the separation. An approximately 5-fold increase in productivity was observed as compared to the conventional batch process [3]. The important observation was that the direct contact of this resin with the microbial cells had no adverse effects on the

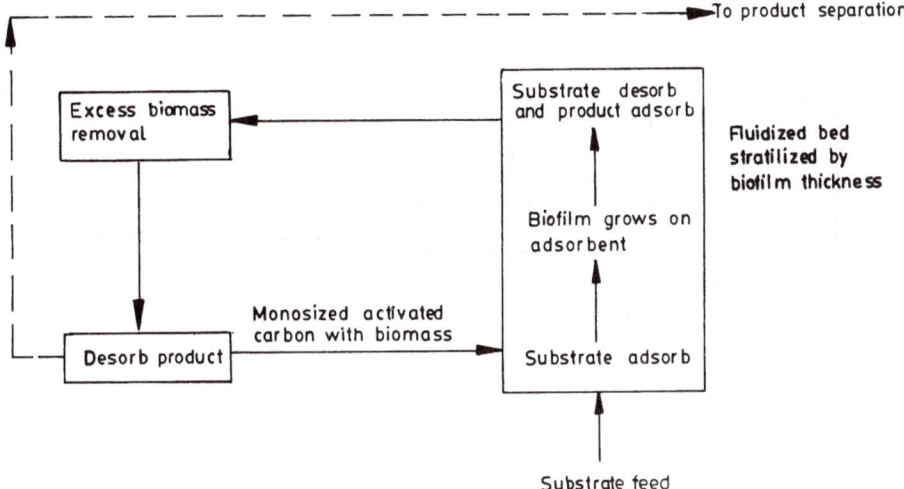

Fig. 3. General arrangement of the extractive bioprocess coupled with ion-exchange [3]

L. delbrueckii cell activity. Using the ion-exchange separation technique, it is reported [50] that a very high lactic acid concentration (265 g l^{-1}) can be achieved in a bioreaction mixture with intermittent feeding of substrate.

Although separation by ion-exchange resins facilitates the rapid removal of ionizable products, the toxicity of the resin to microbes creates a major problem in the case of in-situ product recovery. The ions replaced by the product from resin may act as an inhibitor for microbial growth. Moreover, ionic compounds of a bioreaction mixture and other ionizable metabolites are also adsorbed on to the resins, thus decreasing its product-removing capacity. If the ions being adsorbed happen to be inhibitors, the technique may result in enhanced bioconversion rates. The polar environment around the resin sometimes disturbs the sensitive molecules of the bioconversion medium. In-situ product removal also becomes uneconomical if the resins are costly. However, these problems can be overcome by choosing a suitable resin.

4.3 Liquid-Liquid Extraction

The liquid-liquid extraction for removal of biological products involves either the direct introduction of a solvent into the bioreaction mixture which selectively removes the desired product, or the recycling of the bioreaction mixture itself through an extraction unit in which the desired product is separated by a continuous flow of solvent while the remaining bioreaction mixture is recycled back into the reactor. The general arrangements used in this technique are shown in Fig. 4. For a successful extractive bioconversion process, the solvent should have the following properties. The extraction capacity (K_D) of the solvent should be

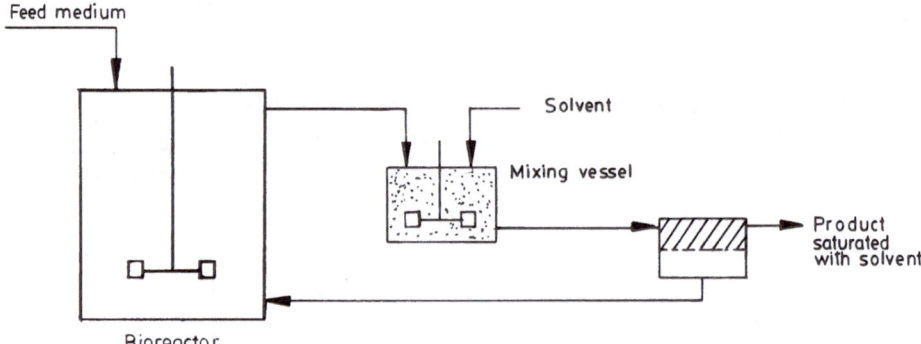

Fig. 4. General arrangement of the extractive bioconversion process coupled with liquid-liquid extraction

very high for the desired product and very low for the essential nutrients. It should have low density, be thermostable and sterilizable. Low cost of solvent and easy separation are also desirable. The screening of the solvent for extractive bioconversion is based not only on the K_D value, but also on its biocompatibility. Suitable solvent selection has been the subject of considerable research [7, 51–57]. Playne and Smith [58] examined thirty suitable organic solvents for their effect on commercial inoculum of facultatively-anaerobic acid-producing bacteria (Table 2). Of these thirty solvents, thirteen were found to be nontoxic to acidogenic bacteria. The presence of the solvent in the bioreaction mixture may adversely affect the catalytic activity of the microbial cells through various mechanisms [59]: (a) as the solvent is transported inside the cells, it combines with cellular components such as enzymes and coenzymes, and interferes in the metabolism; (b) some solvents exhibit a surface-activity property which may affect the permeability of microbial cell membranes with a consequent leakage of metabolites and hindrance to the transport system of microbial cells.

The solvent toxicity to the lactic acid-producing bacteria is on the basis of either polarity or log P values (P is partition coefficient in aqueous system) or the chemical structure of the solvent. According to Brink and Tramper [60], solvents with low polarity and high molecular weight tend to be biocompatible. Laane et al. [61] have given a practical approach for finding the biocompatible solvents. The parameter log P, (where P is the partition coefficient in an n-octanol system) reflect the solvent polarity. The solvents having log $P > 4$ do not appear to have toxic effects on biocatalytic activity. Some of the solvents satisfying this condition have been tested for their effect on *Lactobacillus delbrueckii* (Table 3). Roffler et al. [8] have observed that the solvent toxicity to *L. delbrueckii* increases in the order alkane = cumene < ketone < tertiary amine < secondary amine < quaternary amines. The solvents having a high extraction capacity are found to be more toxic to *Lactobacilli*, and are therefore not suitable for extractive bioconversions involving direct contact between the solvent and the bioreaction mixture. Low K_D-value solvents are not suitable because

Table 2. Effect of solvents on lactic acid bacteria

Solvents	Distribution coefficients (K_D)	Behaviour of solvents
Dipropyl ketone	0.048	Toxic
Isoamyl alcohol	0.447	Toxic
2-ethyl hexanol	–	Toxic
Di-Isopropyl ether	–	Toxic
n-Hexanol	0.313	Toxic
n-Octane	0.195	Toxic
Di-Ethyl ketone	0.164	Toxic
Benzene	< 0.010	Toxic
n-Dodecanol	–	Toxic
Isoamyl acetate	0.107	Toxic
o-Xylene	–	Toxic
Methyl isobutyl ketone	0.113	Toxic
Toluene	< 0.010	Toxic
Nitrobenzene	< 0.010	Toxic
Primene JMT	–	Partially toxic
Amberlite LA-2	–	Partially toxic
n-Hexane	< 0.010	Non-toxic
n-Decane		Non-toxic
Di-butyl phthalate		Non-toxic
Tri-oleyl phosphate		Non-toxic
Tri-octyl phosphate oxide		Non-toxic
Di-isoamyl ether		Non-toxic
Kerosine		Non-toxic
Freon 113		Non-toxic
Di-isoamyl phthalate		Non-toxic
Iso-octane		Non-toxic
Tri-butyl phosphate		Non-toxic
Tri-octyl amine		Non-toxic
Aliquat 336		Non-toxic

Table 3. Toxic effect of solvents on *L. delbrueckii*

Solvents	Log P	Cell viability[a]
Dodecane	6.7	0.97
Isopar M	6.9	0.97
Oleyl alcohol	7.0	1.00
Dodecanol	5.0	0.37

[a] Ratio of the viable cell density to initial cell density after 8 hours contact with the solvent.

of their very low extraction capacity, even if these are nontoxic to microbes. Table 4 summarizes the effect of some of the solvents on the *Lactobacillus* species.

Yabanavar and Wang [7] reported that oleyl alcohol is nontoxic to *L. delbrueckii*. Addition of the amines to the solvent increased the solvent extrac-

Table 4. Effect of solvents on extractive lactic acid bioconversion

Solvent	K_D[a]	% solvent	Toxicity	μ_m	Ref.
Nil			Non-toxic	0.58	64
Heptadecane	0.01	Saturation	Non-toxic	0.55	64
Kerosine	0.6–0.8	Saturation	Partial toxic	0.38	64
Saturated with TOPO		10% saturation	Non-toxic	0.54	64
Tri-butyl phosphate	0.90	Saturation	Toxic	–	64
		10% saturation	Nontoxic	0.65	64
30% (w/w) TOPO in cumene	1–1.4	Saturation	Toxic	–	64
		10% saturation	Non-toxic	0.62	64
Aliquat 336	1–4.5	10% saturation	Toxic	–	64
Iso-butanol	0.59	10% saturation	Toxic	–	64
Liquid paraffin	0.16	Saturation	Non-toxic	0.34	11

[a] Ratio of lactic acid in the solvent to that in the aqueous phase.

tion efficiency but also increased the toxic effect. As a trade off between biocompatibility and extraction efficiency, 15% Alamine-336 in oleyl alcohol was found as the best solvent for extractive lactic acid bioconversion. Using these solvents, extractive lactic acid bioconversion was carried out and only an 18% increase in lactic acid concentration was observed as compared to the conventional batch process. In their studies with immobilized *L. delbrueckii* cells in a Kappa-caregenan support, they observed an increase of 71% in the lactic acid concentration as compared to the control batch process, indicating increased resistance to solvent toxicity on immobilization of cells [9, 10].

Weiser and Geankoplis [62] reported isoamyl alcohol as the best solvent for lactic acid extraction because it is a good selective solvent. Ratchford et al. [63] suggested a useful method for the solvent-extraction of lactic acid, i.e. to neutralize the acid first with organic tertiary amines, thus forming its salts which can then be extracted using chloroform or alcohols. Fillachione and Fisher [64] have given a very different method for lactic acid extraction and purification. The method involves passing gaseous methanol through aqueous lactic acid and condensing the effluent vapours. Methyl lactate is then hydrolyzed to obtain purified lactic acid. These methods are applicable for lactic acid purification but their effect on lactic acid-producing bacteria is still unknown. Extractive lactic acid bioconversion by *Lactobacillus delbrueckii*, using a paraffin oil in conjunction with the solid sorbent (Bonopore), also resulted in an increase in the lactic acid productivity by 14% due to the enhanced extraction efficiency (K_D value) of the pure paraffin oil in presence of solid sorbent [11].

Although some solvents have good selectivity for lactic acid extraction, their direct contact with the bioreaction mixture adversely affects the stability of the microbes. Hence, these solvents cannot be used for such extractive procedures.

4.4 Aqueous Two-phase Systems

Aqueous two-phase systems are generally used in order to overcome the problems connected with the organic liquids in extractive processes. The principle is based on the fact that phase separation occurs when two solutions of water-soluble polymers are mixed (Fig. 5a). The most commonly used polymers are polyethylene glycol (PEG) and Dextran. PEG is a linear synthetic polymer having molecular weight ranging from 200 to 40,000. Dextran is a natural polymer produced by the lactic acid bacterium *Leuconostoc mesenteroides* [12]. It is a polymer of glucose with mainly $< -1,6$ linkages and branched through $< -1,3$ linkages. Dextran has a more hydrophilic character than PEG. Under given conditions, it can be represented by a phase diagram (Fig. 5b) for a two-component system. The tie line relates the mass transfer rates as:

$$\text{mass}_{(top)}/\text{mass}_{(bottom)} = PB/PT$$

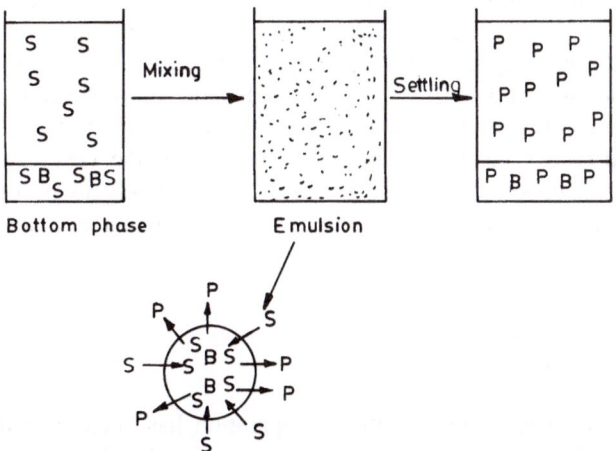

Fig. 5a. Schematic illustration of bioconversions in an aqueous two-phase system [12]. (S = substrate, P = product, B = biocatalyst)

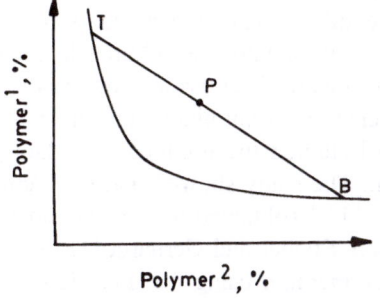

Fig. 5b. Phase diagram for two polymers [12] *1* and *2*, and water. Polymer concentrations represented by points (P) above the binodal curve result in phase separation. Below the curve, the mixtures are homogeneous

In practice, mass ratios are considered as the volume ratios (V_{top}/V_{bottom}). This extractive bioconversion technique has been applied for ethanol, lactic acid and acetone-butanol bioconversion [13, 65–67]. An aqueous two-phase system overcomes some of the problems associated with extractive bioconversion involving liquid-liquid extraction. Instead of adding organic solvents, the polymers are added to the bioreaction mixture until the two separate phases form. Normally, the phase containing more water is biocompatible and the microbes remain there in that phase. The products will get distributed between the two phases depending upon their molecular weights. The phase volumes are adjusted such that the phase containing cells has much smaller volume than the other phase where the cells are absent and that can be drawn off and processed by distillation or other means.

The extractive lactic acid bioconversion by *E. coli*, using an aqueous two-phase system, made up of 6% (w/w) dextran and 2.5% (w/w) PEG, increased the lactic acid productivity by more than 50% over a conventional batch process [13].

Aqueous two-phase separation in extractive bioconversion is still under development. Important variables in a large-scale process include the cost of polymers, its renewal frequency, the ease of product recovery from polymer mixture and the by-product problems during long-term operation. The advantages of aqueous two-phase systems include biocompatibility, possible reduction in waste-water treatment, general applicability to many products of bioconversion and the possibility of concentrating and recovering by-products [8]. Because of the very low K_D values of the polymers, aqueous two-phase systems are not likely to be of practical use for lactic acid production unless a very low cost polymer is available.

4.5 Membrane Separation

Extractive lactic acid bioconversion with membrane separation has been carried out using one of the following extraction techniques: (a) dialysis, (b) electrodialysis, (c) reverse osmosis, and (d) supported liquid membrane separation. In dialysis, electrodialysis and reverse osmosis bioconversion, solid membranes are used for the separations. The dialysis bioconversion system consists of two chambers separated by a solid membrane. In one of the chambers, known as the culture chamber, the bioreaction take place while the other chamber acts as a reservoir. The nutrients from the medium reservoir diffuse into the culture chamber while metabolites and products diffuse into the medium reservoir. Low-product concentrations are maintained in the culture chamber minimizing the effect of metabolite inhibition. Friedman and Gaden [14] studied the production of lactic acid by *L. delbrueckii* in dialysis bioconversion. The study showed that the lactic acid productivity could be increased from 5 g l^{-1} h^{-1} (obtained in a conventional batch) to 8 g l^{-1} h^{-1} using dialysis bioconversion. Stieber and Gerhardt [15] have also studied lactic acid bioconversion in a whey medium using the dialysis technique and obtained a total lactic acid concentration of 88.5 g l^{-1} (52.5 g l^{-1} in a bioreactor and the rest in a dialyzer). They used the continuous dialysis process and recovered the lactic acid as ammonium lactate. Boyaval et al. [16] carried out continuous lactic acid extractive bioconversion using the electrodialysis bioconversion system and obtained a maximum productivity of 22 g l^{-1} h^{-1} at a dilution rate of 0.88 h^{-1}. Yao and Toda [17] have also used the electrodialyzer with a bioreactor. The rate of lactic acid production by the L. delbrueckii strain YPX could be greatly improved by converting the acid to sodium lactate and continuously removing the salt from the bioreaction mixture using an electrodialyzer. Hongo et al. [18] have also applied electrodialysis bioconversion using anion-exchange membranes for alleviating the inhibitory effect of lactic acid in its own bioconversion, and obtained a 3-fold increase in lactic acid productivity as compared to that of a conventional batch process. Schlicher and Cheryan [19] have used the reverse osmosis membrane for concentrating lactic acid and overcoming the problems associated with lactic acid inhibition. Thin film composite membranes made up of cellulose acetate were used for lactic acid removal.

An on-line extractive lactic acid bioconversion system, using a supported liquid membrane, recently termed "pertractive bioconversion", was investigated by Nuchnoi et al. [20] in order to overcome end-product inhibition and consequent enhancement of the productivity of the microbial acidogenesis process. They observed a 5-fold increase in productivity as compared to that of a conventional acidogenic bioconversion without extraction.

The limitations of membrane separation processes are mainly due to the fact that the membranes are expensive and their lifetimes are hard to predict. Dialysis can be applied to any diffusible product but the inherent dilution means that the subsequent recovery cost will increase. Proteins and other compounds of the medium can foul membranes and the accumulation of nondiffusible metabolites will inhibit the cells. The physical stability of the supported liquid membrane is also low and thus cannot be applied to large volumes.

5 Comparison of Extractive Lactic Acid Bioconversion Processes

Extractive lactic acid bioconversion processes are compared either on the basis of extractant properties (Table 5) or on the basis of lactic acid productivity (Table 6).

6 Factors Influencing Extractive Lactic Acid Bioconversion

The parameters affecting the extractive bioconversion of lactic acid are: (a) temperature, (b) pH, (c) substrate, (d) cell modification, and (e) mode of extraction.

Table 5. Extractive lactic-acid bioconversion processes based on extractant properties

Extractive processes	Compatibility	Suitability	Properties of extractant		
			Efficiency	Sterilizability	Cost
EFIRS	Non-toxic or low toxic	For ionisable products only	High because of better selectivity non-specific adsorption limits its capacity	Sterilizable	Economic because reusable cost
EFA	Toxic or retarded	For wide range of products		Sterilizable	can increase due to thermal treatment
EFLE	Toxic	For wide range if direct contact is avoided	Depends on K_D value	Only filter sterilizable	Economic if solvent is not costly
EFATP	Non-toxic	For diffuseable products only	Depends on concentration gradient and product molecular weight		
EFD	Non-toxic	For diffuseable products only	Depends on pressure gradient		High maintenance cost
EFLM	Non-toxic	For diffuseable products only	Depends on K_D value		Economic – depends on volumetric loading capacity

EFIRS : extractive bioconversion using ion-exchange resin for separation.
EFA   : extractive bioconversion using adsorption.
EFLE : extractive bioconversion using liquid-liquid extraction.
EFAT : extractive bioconversion using aqueous two-phase system.
EFD   : extractive bioconversion using dialysis for separation.
EFLM : extractive bioconversion using liquid membrane.

6.1 Temperature

A change of temperature in extractive bioconversion not only affects the process, but also the efficiency of the technique used for lactic acid separation, and hence the degree of lactic acid inhibition. Srivastava et al. [3] have studied the effect of temperature in the range of 37–45 °C on extractive lactic acid bioconversion using an ion-exchange resin. Lowering the temperature causes an increase in the adsorption rate, while higher temperatures in the range of 44–45 °C favour the lactic acid bioconversion rate. The combined effect of these two rates determines the productivity in extractive bioconversion. They obtained an optimum temperature of 39°C for extractive lactic acid bioconversion. In some of the extractive lactic acid bioconversion processes, the extraction is carried out at lower temperature than that of the bioconversion in order to facilitate the two temperature-dependent processes. In those cases, the cells are not allowed to undergo temperature shock. They are first separated from the bioreaction mixture, then mixed again with the

Table 6. Comparison of different extractive lactic acid bioconversion processes on the basis of their effect on lactic acid

Reaction systems	Productivities		References	Remarks
	Lactic acid	Cell mass		
Ultrafiltration: with electrodialysis without electrodialysis	85 ± 5^a 64^b	22^b	Boyaval et al. [16]	
Dialysis Without dialysis	8^a 5^a		Friedmann and Gaden [14]	
Ultrafiltration and electrodialysis	82.2^b		Hongo et al. [18]	Although 5.5 times productivity but membrane fouling, restricted application
Ultrafiltration and electrodialysis			Roucourt et al. [68]	Develop a model, obtained $\mu_m = 1$ for *S. thermophilus*
Electrodialysis	8.67^a	10.5^b	Yao and Toda [17]	Obtained 90^b lactic acid concentration in permeate side
Dialysis	88.5^b		Stieber and Gerhardt [25]	Total 88.5^c, the 52.5^c in bioreactor and 36^c in dialysate
Liquid membrane	5 fold compared to batch		Nuchnoi et al. [20]	Stability of liquid membrane is very low
Reverse osmosis			Schlicher and Cheryan [19]	Concentrated recovery of product by the continuous process
Liquid-liquid extraction	12^c		Yabanavar and Wang [9,10]	Many solvents tested, a 15% alanine in oleyl alcohol was found to be based for extractive lactic acid bioconversion
Ion exchange resin	5.32 fold compared to batch		Srivastava et al. [3]	Batch recycle system was used
Aqueous two-phase system	50% more than conventional batch		Hahn–Hagerdal et al. [67]	*E. coli* was used for lactic acid production

[a] $g\ l^{-1}h^{-1}$, [b] $g\ l^{-1}$ and [c] g per l gel per h.

bioreaction mixture after the separation of the lactic acid. Various research papers have reported such techniques [22, 23, 68]. Both the bioreactor and the cell separation units are maintained at the same temperature while the extraction unit is maintained at a lower temperature.

6.2 pH

The control of pH in extractive bioconversion is very critical. The two simultaneously occurring processes, namely, the bioconversion and the separation by
extractants such as liquid solvents, adsorbents, ion-exchangers and membranes,
are pH-dependent. The increase in the extracellular H^+ ion concentration during
production of lactic acid may cause several problems such as: (a) disruption of
the plasma membrane of the cell, (b) cell enzyme inhibition, (c) alteration of the
ionization of the nutrient molecules and reduced nutrient availability to the cells,
and (d) a restricted substrate uptake rate. The pK_a value for lactic acid is 3.8.
The pH for extractive lactic acid biconversion is usually in the range 4.0–7.0.
For separation processes dependent on the ionic form of product, the pH is set
in the range of 5.0–7.0, which is far above the pK_a value. Therefore, ionization
of the lactic acid would substantially decrease if the pH decreased below 7.0.
Consequently, the rate of separation will drop if it is favoured by an increased
ionization of the lactic acid, as observed in the case of ion-exchange separation
[3]. On the other hand, the separation rate will increase if it is favoured by the
unionized form of lactic acid, as observed in the case of liquid-liquid extraction [7,9,10] and simple adsorption [42,43] where the pH is set in the range of
4.0–5.0.

Yabanavar and Wang [7,9,10] used extractive bioconversion with liquid-liquid
extraction. They observed a 217% increase in cell density when the pH is decreased from 4.5 to 4.2. Nuchnoi et al. [20] have investigated the pH effect on
extractive lactic acid bioconversion using a supported liquid membrane. They used
a pH of 4.8 which is lower than the growth optima of the acidogenic bacteria
for acidogenic bioconversion, and obtained a 3 times greater yield of acid in the
bioreaction as compared to the conventional batch process.

6.3 Substrate

Many industrial raw materials such as potato [69], whey [70–76], corn [77],
soymilk [78] and sulphite waste liquor [79] have been employed for lactic acid
production. It has been observed that potato [69], without supplementing other
nutrients in the medium, can produce a maximum lactic acid yield of 0.95 $g\,g^{-1}$
(of sugar) in a batch process. Tuli et al. [75] observed a maximum lactic acid productivity of 0.58 $g\,l^{-1}\,h^{-1}$ by immobilized *L. casei* in a whey permeate medium
containing 15 $g\,l^{-1}$ of yeast extract. Aeschlimann and von Stockar [70–72] also
carried out the lactic acid bioconversion in a chemostat by using a whey medium
mixed with 10 $g\,l^{-1}$ of yeast extract. Using two in series, they achieved a lactic
acid productivity of 6.4 $g\,l^{-1}\,h^{-1}$ at a dilution rate of 0.2 h^{-1}, while in a single stage at a very low dilution rate of 0.06 h^{-1}, a lactic acid productivity of
8.3 $g\,l^{-1}\,h^{-1}$ was observed. In a batch process, a specific lactic acid productivity
of 0.027 h^{-1} by *L. diaacetolactis* grown on soymilk, was reported by Patel et al.
[78]. Sulphite waste liquor, corresponding to 1 ton of pulp derived from spruce

wood, yielded 129.4 kg of lactic acid [79] in a batch process. Inskeep et al. [77] were also able to produce 180 g l^{-1} lactic acid from corn sugars in a batch process. Srivastava et al. [50] carried out extractive lactic acid bioconversion using a molasses medium. Using an intermittent C-substrate feeding system, they obtained a lactic acid productivity of 0.94 g l^{-1} h^{-1} and achieved a very high lactic acid concentration of 265 g l^{-1} in the bioreaction mixture.

6.4 Cell Modification

The end-product inhibition observed in lactic acid bioconversion results in low process productivity. Higher productivity can be achieved either by adopting new bioconversion techniques or by improving the resistance of the microorganisms and thus enhance their capacity to grow and produce lactic acid at even higher concentrations. The microbial resistance to lactic acid concentration can be developed either by mutation, or by physical changes in the microorganisms. Cell adaptation is also one of these methods. Srivastava et al. [50] found that the combined effect of extraction and the use of lactic acid-resistant cells can lead to a very high lactic acid concentration. They also observed that the size of the *Lactobacillus delbrueckii* cells was significantly reduced by a increase in the lactic acid concentration in the bioreaction mixture. The cell dimension was reduced from 1.17 μm×6.11 μm to 0.46 μm×1.31 μm when the extracellular lactic acid concentration gradually increased from 20 to 265 g l^{-1}. No adverse effect on the cell activity was observed. The exact phenomenon is not yet known, but the observation gives a clear indication of some relationship between the adaptation of the cells to high lactic acid concentration and cell morphology.

6.5 Mode of Extraction

The mode of extraction is also very important for the extractive lactic acid bioconversion process. A continuous or semicontinuous mode of extraction is generally used in extractive lactic acid bioconversion processes. Extractive bioconversion processes involving ion exchange for separation are carried out in a semi-continuous manner [3] because this technique enables lactic acid to be removed at a fast rate. This mode of extraction helps in maintaining the pH of the reaction mixture in bioreactor and eliminates the need for alkali to be added in order to regulate the pH. If operated in the continuous mode, the removal of lactic acid and the input of the replaced anion from the ion exchange resin to the reaction mixture may cause a drastic pH increase if the lactic acid production rate is lower than that of its continuous separation rate. This may result in a pH shock to the bacteria that produce the lactic acid unless the pH in the bioreactor is controlled within limits.

In the case of the extractive bioconversion technique that uses adsorbents for separating the lactic acid, a semi-continuous mode of extraction was adopted

[6]. The increase in lactic acid concentration inside the fluidized-bed bioreactor caused a decrease in pH which was controlled by the addition of polymeric resin from the top of the bioreactor in a semi-continuous manner. The lactic acid adsorbed on the resin particles left from the bottom of the bioreactor. Yabanavar and Wang [9, 10] also controlled the pH of the bioreaction mixture in their extractive bioconversion process involving liquid-liquid extraction in a semi-continuous manner. The pH controller monitored the pH of the reaction mixture inside the bioreactor and accordingly activated the inlet and exit pumps to initiate the extraction operation.

However, in the case of extractive bioconversion involving membrane separations, the continuous mode of extraction is generally adopted for lactic acid in-situ removal as the technique provides a slow removal of lactic acid from the bioreaction mixture [15–20]. Some of the extractants are toxic to microorganisms; in these cases direct contact between the microorganism and the extractant should be avoided and the cells must be separated from the bioreaction mixture before treating it with the extractant [9, 15, 19].

7 Mathematical Models

7.1 The Friedman–Gaden Model

A model depicting cellular growth and lactic acid production has been proposed by Friedman and Gaden [14]. This model is a modification of the model proposed by Luedeking and Piret [80] and incorporates the lactate inhibition effects. It relates the lactic acid production rate and the growth rate in the following manner:

$$[1 - (P(t)/P_{\max})] = \alpha(P)[1 - (N(t)/N_{\max})] \tag{7.1.1}$$

The dialysis culture system has been modelled as shown in Figs. 6a and 6b where D represents the lactic acid transferred by dialysis to a reservoir. Using mass

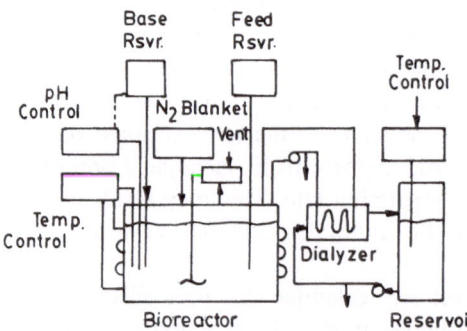

Fig. 6a. Schematic diagram of the dialysis bioconversion system [14]

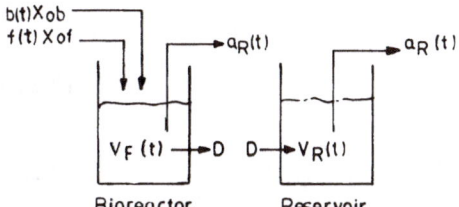

Fig. 6b. Representation of the dialysis bioconversion system for modelling [14]

balance, they have developed the following expression for the specific growth
rate:

$$N(t) = (1/n) \, (dn/dt) + (f + b)/V \tag{7.1.2}$$

where f and b are the volumetric flow rates of nutrient and base into the bioreactor respectively; n is the cell concentration, $N(t)$ represents the specific growth
rate of cells, $P(t)$ is the specific product formation rate, and $\alpha(P)$ is an empirical
model constant.

7.2 The Stieber–Gerhardt Model

The previous model correlated the specific growth rate with the specific lactic acid
production rate. Although, this model is in good agreement with the experimental
results, it describes a bioconversion system involving very low substrate and
product concentrations. The terms in the model have no biological significance.
Keeping this in view, Stieber and Gerhardt [21–25] proposed a model based on
substrate-limitation and product-inhibition terms which have a biological basis.
A mechanism has been proposed for the model. Once the substrate is taken up
by the cells, it is metabolized irreversibly to lactic acid, which then inhibits the
cells. They have assumed that there is an adequate concentration of the cells and
the substrate, and that the rates of substrate utilization and product formation
depend mainly on the concentration of the product in the environment around
the cells. The model parameters have been estimated from the steady-state data.
The major shortcoming of the model is that the product-inhibition effects are not
strong enough since the lactic acid concentration is below 70 g l^{-1}.

The rate equations have been combined with the material balance equations
and the variables defined in the form of dimensionless parameters to obtain a
generalized model for bioconversion. The resulting equations for various circuits
in the bioreactor are given as follows:

$$\frac{d\overline{P}_r}{dt} = \left(-(1 + R\pi)\overline{P}_f + \left[\theta \overline{S}_f / \left(\overline{K}_s + \overline{S}_f \overline{K}_p \overline{P}_f \right) + \beta T_f / \alpha_1 \right] \overline{X}_f \right.$$
$$\left. + R\pi \overline{P}_d + \overline{P}_f^0 \right) / T_f \tag{7.2.1}$$

$$\frac{d\bar{S}_f}{dt} = \left[-(1+\pi)\bar{S}_f - \left[\theta\left(\bar{S}_f/\left(\bar{K}_s + \bar{S}_f + \bar{K}_p\bar{P}_f\right)\right) + \beta T_f/\alpha_1\right]\bar{X}_f$$
$$+ \pi\bar{S}_d + 1\right]/T_f \tag{7.2.2}$$

$$\frac{d\bar{X}_f}{dt} = \left[\theta\left(\bar{S}_f/\left(\bar{K}_s + \bar{S}_f + \bar{K}_p\bar{P}_f\right)\right) - 1\right]\left[\bar{X}_f/T_f\right] \tag{7.2.3}$$

Similarly for dialysate circuits, the equations are

$$\frac{d\bar{S}_d}{dt} = \left[-(\phi + \pi)\bar{S}_d + \pi\bar{S}_f\right]\left[F_f/V_d\right] \tag{7.2.4}$$

$$\frac{d\bar{P}_d}{dt} = \left[-(\phi + R\pi)\bar{P}_d + R\pi\bar{P}_f\right]\left[F_f/V_d\right] \tag{7.2.5}$$

where

$\bar{P}_d = P_d/\Gamma\ S_f^0$	i.e., product factor in the circuit of dialysate,
$\bar{P}_f = P_f/\Gamma\ S_f^0$	i.e., product factor in the circuit of bioreactor,
$\bar{P}_f^0 = P_f^0/\Gamma\ S_f^0$	i.e., product factor in the circuit of bioreactor feed,
$\bar{S}_d = S_d/S_f^0$	i.e., substrate factor in the circuit of dialysate,
$\bar{S}_f = S_f/S_f^0$	i.e., substrate factor in the circuit of bioreactor,
$\bar{X}_f = X_f/S_f^0$	i.e., cell factor in the circuit of bioreactor,
$R = P_{mp}/P_{ms}$	i.e., ratio of product to substrate membrane permeabilities,
$\pi = P_{ms}A_m/F_f$	i.e., membrane permeability factor,
$\phi = F_d/F_s$	i.e., feed flow ratio,
$\bar{K}_s = K_s/S_f^0$	i.e., substrate limitation factor
$\bar{K}_p = \Gamma\ K_p$	i.e., product inhibition factor,
$\theta = \mu_m\ T_f$	i.e., time factor.

These mathematics showed the permeability factor to be as low as 0.4. Consequently, the bioconversion could be improved considerably by using more permeable membranes. Increased conversion of substrate could be made possible by increasing the cell mass concentration in the dialysis bioreactor. An ideal configuration for maximum productivity would be a batch reactor followed by a dialysis bioreactor.

7.3 The Yabanavar–Wang Model

Yabanavar and Wang [10] have developed a process model for extractive lactic acid bioconversion using liquid-liquid extraction. Basically the process has been modelled so as to interrelate the various process parameters such as flow rates, concentration, pH, etc. The bioconversion system is shown schematically in the Figs. 7a and 7b. The material balance for lactic acid around the bioreactor can

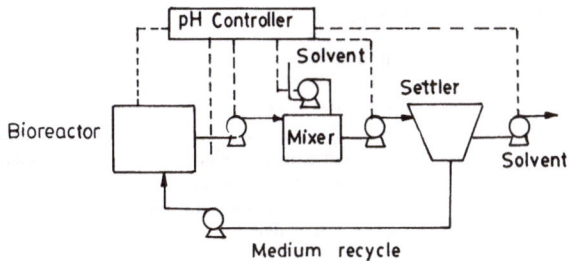

Fig. 7a. Schematic diagram of the extractive bioconversion system using liquid-liquid extraction [9, 10]

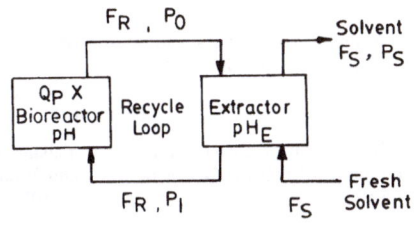

Fig. 7b. Flow stream designation for extractive bioconversion [9, 10]

be written as

$$F_R P_0 = F_R P_1 + Q_P X V \ . \tag{7.3.1}$$

From the above equation, we can write

$$D_R P_0 = D_R P_1 + Q_P X \tag{7.3.2}$$

where $D_R = F_R/V$. The lactic acid balance around the extractor can be written as

$$F_R(P_0 - P_1) = F_S P_S \ . \tag{7.3.3}$$

Putting $D_S = F_S/V$

$$D_R(P_0 - P_1) = D_S P_S \ . \tag{7.3.4}$$

Combining equations (7.3.2) and (7.3.4) they obtained

$$D_S P_S = Q_P X \tag{7.3.5}$$

Lactate is present at a high concentrations in the medium compared to other anions. Its charge is balanced by counterions, such as potassium, that are present in the medium. As these cations are not removed by extraction, it is expected that, because of the charge balance, the lactate concentration should remain unchanged during the extraction process. Therefore, the lactate concentration in the bioreactor and the extractor could be written as follows:

$$P_0/P_1 = (1 + 10^{pK_a - pH_e})(1 + 10^{pK_a - pH})^{-1} \tag{7.3.6}$$

Table 7. Different models for extractive lactic acid bioconversion

Extractive techniques	Features	Limitations	Authors
Dialysis bioconversion	1. Modified Luedeking–Piret model is proposed for lactic acid bioconversion 2 Lactate inhibition effects are incorporated 3. The model is modified by taking into account that at low lactate concentration, the acid production rate is independent of growth rate	Model describes a system involving a low product concentration. The model is based upon the correlation with the experimental data	Friedmann and Gaden [14]
Dialysis bioconversion	1. Model contains substrate and product limitation terms which have a biological basis	The model does not state the strong product inhibition effect ($> 70 \, g \, l^{-1}$)	Stieber and Gerhardt [21]
Extractive bioconversion using liquid-liquid extraction	1. Designed the process model for extractive bioconversion using liquid-liquid extraction	Solvent diffusion through the immobilized cell beads can be a limiting factor	Yabanavar and Wang [9, 10]

Since undissociated acid is the extractable species, the pH in the extractor, pH_E, should be maintained as low as possible for achieving the most efficient operation. It can be observed that pH_E will be low when the recycle rate D_R is large. At a very high recycle rate, the pH in the extractor approaches the pH in the bioreactor. This descriptive process model is helpful in the design and operation of the continuous extractive bioconversion system. The model showed that the solvent requirements can be minimized by using a solvent with a high distribution coefficient and maintaining a moderately high product concentration in the bioreactor. The basic disadvantage of this model is that it does not take into account lactic acid inhibition and the substrate limitation effects. Moreover, this model is unstructured and therefore cannot be used to explain the mechanism of bioconversion.

The limitations and features of these models have been comprehensively compared in Table 7.

8 Conclusions and Future Scope

The extractive bioconversion of lactic acid can only employ separation techniques such as liquid-liquid extraction, ion-exchange separation, adsorption on polymeric resin and membrane separations. So far, efforts have been directed mainly towards using either liquid-liquid extraction or membrane separations. As regards liquid-liquid extraction, no potential solvent is available for which there is a good trade-off between its extraction efficiency and its toxicity towards microbes in those extractive bioconversions where direct contact between the solvent and the bioreaction mixture is allowed. The membrane separation techniques involving reverse osmosis, dialysis, electrodialysis and liquid membranes work well for extractive bioconversion of lactic acid. The use of selective membranes for the process can increase the efficiency. However, some other factors such as the fouling of the membrane and its transport capacity, its own cost, and the cost of maintaining a constant driving force across the membrane limits the application of this technique in industry. Thus, extractive lactic acid bioconversion involving liquid-liquid extraction or membrane separations may be unattractive for industrial bioconversion.

Although, extractive lactic acid bioconversions involving ion-exchange in a resin-packed separating column [3] lead to a very high concentration of lactic acid in the reaction mixture, the productivity of lactic acid in this process is nevertheless unsatisfactory. Separation of lactic acid at high concentration may be more attractive from the point of view of recovery cost. A column packed with a suitable ion-exchange resin (such as Amberlite IRA-400, OH^--form) gives high separation rates, has less fastidious maintenance and operational requirements, and is biocompatible. Moreover, such systems help in cell adaptation. A continuous culture with the adapted cells may increase the process productivity substantially. In fact, a successful process should include the advantages of both higher productivity and lower recovery costs. All these advantages are not found at the same time in other systems. This makes such a system most promising and attractive for extractive lactic acid bioconversion. Some resins are expensive but their reusability offsets this disadvantage.

Extractive bioconversion systems are designed to resolve the problems of end-product inhibition. These systems work very efficiently for alcohols, carboxylic acids, volatile acids and antibiotic bioconversions without affecting the physico-chemical nature of the product. They permit higher productivities while at the same time reducing the requirements of waste-water treatment, demanded by the use of concentrated feed stock. However, these advantages are partially offset by the complexities of the process. Although most studies involving in-situ product recovery use batch or semicontinuous operations, the advantages of in-situ product recovery may be more pronounced with continuous systems because of the inherent advantages of such systems over other layouts. An effort in this direction is very much needed as regards the application of such processes in industry.

9 References

1. Blondeau L (1947) J Pharmacol 12: 244
2. Vickroy TB (1981) In: Young MM, Robinson CW, Vezina C (eds) Comprehensive biotechnology, vol 1. Pergamon, New York, p 761
3. Srivastava A, Roychoudhury PK, Sahai V (1992) Biotechnol Bioeng 39: 607

4. Wang HY, Kominek LA, Jost JL (1980) In: Young MM (ed) Adv Biotechnol 1: 601
5. Wang HY (1983) Ann N Y Acad Sci 413: 313
6. Davison BH, Thompson JE, Scott CD (1990) Abstr Paper Amer Chem Soc 200 meet, Pt 1BIOT93
7. Yabanavar VM, Wang DIC (1985) Ann N Y Acad Sci 523:
8. Roffler SR, Blanch HW, Wilke CR (1984) Trends in Biotechnol 2: 129
9. Yabanavar VM, Wang DIC (1991) Biotechnol Bioeng 37: 1095
10. Yabanavar VM, Wang DIC (1991) Biotechnol Bioeng 37: 544
11. Seevaratnam S, Holst JO, Hjorleifsdotter S, Mattiasson B (1991) Bioprocess Eng 6: 35
12. Anderson E, Hahn-Hagerdal B (1990) Enzyme Microb Technol 12: 242
13. Mattiasson B, Hahn-Hagerdal B (1983) In: Immobilized cells and organells, 1, CRC Press, Boca Raton, Florida, p 121
14. Friedman MR, Gaden EL (1970) Biotechnol Bioeng 12: 961
15. Stieber RW, Gerhardt P (1981) Biotechnol Bioeng 23: 535
16. Boyaval P, Corre C, Terre S (1987) Biotech Lett 9: 207
17. Yao PX, Toda K (1990) J Gen Appl Microbiol 36: 111
18. Hongo M, Nomura Y, Iwahara M (1986) Appl Environ Microbiol 52: 314
19. Schlicher LR, Cheryan M (1991) J Chem Technol Biotechnol 49: 129
20. Nuchnoi P, Izawa I, Nishio N, Nagai S (1987) J Ferment Technol 65: 699
21. Stieber RW, Gerhardt P (1979) Appl Environ Microbiol 37: 487
22. Stieber RW, Gerhardt P (1980) J Diary Sci 63: 722
23. Stieber RW, Gerhardt P (1981) Biotechnol Bioeng 23: 523
24. Stieber RW, Gerhardt P (1977) Appl Environ Microbiol 34: 725
25. Stieber RW, Gerhardt P (1977) Appl Environ Microbiol 34: 733
26. Lipinsky ES (1981) Science 212: 1465
27. Stryer L (1988) In: Biochemistry, W H Freeman, New York, p 356
28. Lehninger AL (1982) In: Principles of Biochemistry, Worth, USA, p 414
29. Cordon TC, Treadway RH, Walsh MD, Osborne MF (1950) Ind Eng Chem 42: 1833
30. Prescott SC, Dunn CG (1959) Microbiology, Mc Graw Hill, New York, p 185
31. Stanier RY, Adelberg EA, Ingraham JL (1976) In: The microbial world, Prentice Hall, Englewood Cliffs., New Jersey, p 678
32. Gasser F (1970) J Gen Microbiol 62: 223
33. Leonard RH, Peterson WH, Johnson MJ (1948) Ind Eng Chem 40: 57
34. Nakamura LK and Crowell CD (1979) Dev Ind Microbiol 20: 531
35. Childs CG, Welsby B (1966) Process Biochem 1: 441
36. Viniegra-Gonzelez G, Gomez J (1984) In: Wise D (ed) Bioconversion System, CRC Press, Florida, p 17
37. Koser SA (1968) In: Charles E Thomas (ed) Lactobacillus, Springfield, Illinois, p 340
38. Aborhey S, Williamson D (1977) J Gen Appl Microbiol 23: 7
39. Wang HY, Robinson FM, Lee SS (1981) Biotechnol Bioeng Symp 11: 555
40. Lee SS, Wang HY (1982) Biotechnol Bioeng Symp 12: 221
41. Lenckii RW, Robinson CW and Young MM (1983) Biotechnol Bioeng Symp 13: 617
42. Dykstra KH, Li XM, Wang HY (1988) Biotechnol Bioeng 32: 356
43. Andrews GF, Fonta JP (1989) Appl Biochem Biotechnol 20121: 375
44. Larson M, Holst O, Mattiasson B (1984) Europ Congr Biotechnol 2: 313
45. Finn RK (1966) J Ferment Technol 44: 305
46. Levy PF, Sanderson JE, Wise DL (1981) Biotechnol Bioeng Symp 11: 239
47. Kitai A, Tone H, Ishikura J, Ozaki A (1968) J Ferment Technol 46: 442
48. Tangu SK, Ghose TK (1981) Process Biochemistry Aug/Sep: 24
49. Denkwalter RG, Gillin J (1959) Auslegeschrift, 102891
50. Srivastava A, Roychoudhury PK, Sahai V (1995) Communicated
51. Datta R (1981) Biotechnol Bioeng 23: 61
52. Wang HY, Kominek LA, Jost JL (1980) Adv Biotechnol 1: 601
53. Roddy JW (1981) Ind Eng Chem Proc Des Dev 20: 104
54. Leeper SA, Wankat PC (1982) Ind Eng Chem Proc Des Dev 21: 331
55. Minier M, Goma G (1987) Biotechnol Bioeng 24: 1565
56. Murphy TK, Blanch HW, Wilke CR (1982) Process Biochem 17: 6
57. Murphy TK, Blanch HW, Wilke CR (1982) Process Biochem 17: 40
58. Playne MJ, Smith BR (1983) Biotechnol Bioeng 25: 1251
59. Lamane C, Mallette MP (1965) Chemical Disinfectant In: Basic biotechnology, Williams and Wilkins, Batimore, p 921

60. Brink LS, Tramper J (1985) Biotechnol Bioeng 27: 1258
61. Laane C, Boeren S, Vos K (1985) Trends in Biotechnol 4: 167
62. Weiser RB, Geankoplis CJ (1955) Ind Eng Chem 47: 858
63. Ratchford WP, Harris EM, Fisher CH, Willits CO (1951) Ind Eng Chem 43: 778
64. Fillachione EM, Fisher CH (1944) Ind Engg Chem 36: 223
65. Puzis M, Heden CG (1965) Biotechnol Bioeng 7: 335
66. Kuhn I (1980) Biotechnol Bioeng 22: 2393
67. Hahn-Hagerdal B, Mattiasson B, Albertson PA (1981) Biotech Lett 3: 53
68. Roucourt AD, Girard D, Prigent Y, Boyaval P (1989) App Microbiol Biotechnol 30: 528
69. Cordon TC, Treadway RH, Walsh MD, Osborne MP (1950) Ind Eng Chem 42: 1833
70. Aeschlimann A, von Stockar U (1990) Appl Microbiol Biotechnol 32: 398
71. Aeschlimann A, von Stockar U (1989) Biotech Lett 11: 195
72. Aeschlimann A, von Stasi LD, von Stockar U (1990) Enzyme Microb Technol 12: 926
73. Roy D, Goulet J, LeDuy A (1986) Appl Microb Technol 24: 206
74. Roy D, LeDuy A, Goulet J (1987) Can J Chem Engg 65: 597
75. Tuli A, Sethi RP, Khanna PK, Marwaha SS (1985) Enzyme Microb Technol 7: 164
76. Mahaia MA, Cheryan M (1987) Proc Biochem Dec: 185
77. Inskeep GC, Taylor GG, Brietzke WC (1952) Ind Engg Chem Sep: 1955
78. Patel AA, Waghmare MM, Gupta SK (1980) Proces Biochem Oct/Nov: 9
79. Leonard RH, Peterson WH, Johnson MJ (1949) Ind Eng Chem 40: 57
80. Luedeking R, Piret EL (1959) J Biochem Microb Technol Engg 1: 393

Production of Rhamnolipid Biosurfactants

Urs A. Ochsner[1], Thomas Hembach[2] and Armin Fiechter[3]
[1] Institute for Biotechnology, Swiss Federal Institute of Technology, ETH-Hönggerberg, CH-8093 Zürich, Switzerland. To whom correspondence should be addressed. Present address: University of Colorado Health Sciences Center, Dept. of Microbiology, Campus Box B175, 4200 E Ninth Avenue, Denver CO 80262, USA
[2] Fraunhofer-Institut für Grenzflächen- und Bioverfahrenstechnik, Nobelstr. 12, D-70569 Stuttgart, Germany. Present address: Chema Balcke-Dürr Verfahrenstechnik GmbH, Arnstädter Strasse 22, D-99334 Rudisleben, Germany
[3] Institute for Biotechnology, Swiss Federal Institute of Technology, ETH-Hönggerberg, CH-8093 Zürich, Switzerland

Dedicated to Prof. Dr. J. Klein, Braunschweig, on his 60th birthday in recognition of his outstanding efforts in promoting biotechnology in Germany

Advances in Biochemical Engineering/
Biotechnology, Vol. 53
Managing Editor: A. Fiechter
© Springer-Verlag Berlin Heidelberg 1995

Biosurfactants are of increasing interest due to their broad range of potential applications. A large variety of microbial surfactants is known at present, some of which may be used for specific applications. Towards the large scale industrial production of biosurfactants, the physiology, biochemistry and genetics of biosurfactant synthesis has to be well understood. A fully integrated process has to be developed, allowing high productivities under optimized conditions. In the past few years, we have investigated the molecular biology of rhamnolipid synthesis in *Pseudomonas aeruginosa*. The key enzymes catalyzing the final steps of rhamnolipid biosynthesis have been partially purified and characterized. The structural and regulatory genes encoding the rhamnolipid synthesis pathway have been isolated and characterized. The knowledge of the complex mechanisms involved in rhamnolipid synthesis facilitates the overproduction of these extracellular compounds. Furthermore, the transfer of the relevant genes into other species allows the production of rhamnolipids in heterologous hosts under controlled conditions.

An integrated process for the production of rhamnolipids on an industrial scale has been developed. This process involves continuous cultivation under optimized media and growth conditions and makes use of refined methods of cell recycling, gas exchange and downstream processing, thus allowing high yields and productivities.

List of Symbols and Abbreviations

cmc	critical micelle concentration
CSTR	continuous stirred tank reactor
CTAB	cetyltrimethylammonium bromide
DEAE	diethylaminoethyl
G.R.A.S.	generally regarded as safe
GS-medium	rhamnolipid production medium
IPTG	isopropylthiogalactoside
IS50	insertion element of transposon Tn5
kb	kilobase
lacZ	gene encoding β-galactosidase
LB	Luria broth
LPS	lipopolysaccharide
MEOR	microbial-enhanced oil recovery
Miller units	β-galactosidase activity on the hydrolysis of o-nitrophenyl-β-D-galactoside, $1000 \times A420 \times A600^{-1} \times ml^{-1} \times min^{-1}$
nt	nucleotide
OD	optical density
ORF	open reading frame
pI	isoelectric point
rhamnolipid 1	L-rhamnosyl-L-rhamnosyl-β-hydroxydecanoyl-β-hydroxydecanoate
rhamnolipid 2	L-rhamnosyl-β-hydroxydecanoyl-β-hydroxydecanoate
rhl	designation for genes involved in rhamnolipid production
tac	*trp-lac* fusion promoter inducible with IPTG
TDP	thymidine-diphosphate
TDPG	thymidine-diphospho-glucose
TDPR	thymidine-diphospho-rhamnose
TLC	thin-layer chromatography
Tn5-GM	transposon Tn5 carrying a gentamicin resistance marker
σ^{54}	alternative sigma factor of RNA polymerase

1 Introduction

Microbial surfactants include a structurally diverse group of compounds containing a hydrophilic and a lipophilic moiety within the same molecule. The common lipophilic part is built up by the hydrocarbon chain of a fatty acid, while the polar or hydrophilic group is derived from the ester or alcohol groups of neutral lipids, the carboxylate group of fatty acids or amino acids, the phosphate containing portions of phospholipids, or the carbohydrates of glycolipids. The structures and physico-chemical properties of a large variety of microbial surface-active compounds produced by bacteria, yeasts and fungi have been described in several reviews [1–7]. The best characterized biosurfactants are presented in Table 1.

Due to their amphiphilic character, surfactants partition preferentially at the interface between phases of different degrees of polarity and hydrogen bonding, such as oil/water, air/water or solid/water interfaces. The critical micelle concentration (CMC) is defined by the solubility of a surfactant within an aqueous phase. At concentrations above the CMC, amphiphilic molecules associate readily to form supramolecular structures such as micelles, bilayers and vesicles. Both synthetic or microbial surfactants can lower the air/water surface tension for distilled water from approximately $72 \, \text{mN m}^{-1}$ to roughly $30 \, \text{mN m}^{-1}$. Similarly, the hexadecane/water interfacial tension is reduced from $40 \, \text{mN m}^{-1}$ to $1 \, \text{mN m}^{-1}$ or even less [8].

The physiological roles of biosurfactants are the emulsification of water-insoluble substrates [9] as well as the exertion of antimicrobial effects against competing microorganisms [10].

Biosurfactants are of industrial interest due to their broad range of potential applications including emulsification, phase separation, wetting, foaming, solubilization, emulsion stabilization, de-emulsification, corrosion-inhibition and viscosity reduction. Substances for such applications are needed in agriculture, building and construction, elastomer and plastics production, food and beverage stabilization, in cosmetics and in the cleaning, leather, metal and paper industries. Surfactants are also used in paints and protective coatings, for processing of petrochemical products, in the textile industry and for depollution [11]. The initial focus and most of the industrial interest in biosurfactants has been toward applications for microbial-enhanced oil recovery (MEOR) and various other uses in the oil service industry [12]. Other areas where biosurfactants can successfully be used is the control of oil spills and the microbial decontamination of soil [13–15]. Furthermore, biosurfactants are of increasing importance in the cosmetics industry where compounds of low antiirritating effects and high compatibility with skin are needed [16]. Biosurfactants, it is assumed will play a very important role in the food industry where emulsions of oil or fat dispersed in a water phase have to be formed or stabilized [17].

Table 1. Microbial biosurfactants

Group	Microorganism	Biosurfactant/Bioemulsifier
Glycolipids	*Pseudomonas* sp.	Rhamnolipids
	Torulopsis sp.	Sophorose lipids
	Arthrobacter sp.	Trehalose lipids, sucrose lipids, fructose lipids
	Nocardia corynebacteroides	Penta- and disaccharide lipids Trehalose dimycolic acid
	Serratia rubidaea	Rubiwettins
	Mycobacterium tuberculosis	Trehalose lipids
	Mycobacterium leprae	Phenolic glycolipid
	Lactobacillus fermenti	Diglycosyl diglycerides
	Candida sp.	Mannosyl erythritol lipid
	Candida bogoriensis	Sophorose lipids
	Rhodococcus erythropolis	Trehalose dimycolic acid
	Rhodococcus aurantiacus	Several glycolipids
	Ustilago zeae, U. maydis	Cellobiose lipids
	Shizouella melanogramma	Mannosylerythritol lipid
Polar and neutral lipids	*Corynebacterium lepus*	Fatty acids
	C. alkanolyticum	Lecithin, phospholipids
	Nocardia erythropolis	Fatty acids, neutral lipids
	Arthrobacter parrifineus	Fatty acids
	Talaromyces trachyspermus	Fatty acids
	Rhodotorula sp.	Polyol lipids
	Rhodococcus erythropolis	Phosphatidylethanolamine
	Thiobacillus thiooxidans	Phospholipids
	Capnocytophaga sp.	Sulfonolipids
Amino acid-containing biosurfactants (lipoproteins, lipopeptides, peptidoglyco-lipids, protein emulsifiers)	*Pseudomonas rubescens*	Ornithine-containing lipid
	Pseudomonas fluorescens	Viscosin
	Serratia marcescens	Serratamolide, serrawettins
	Thiobacillus thiooiydans	Ornithine-containing lipid
	Gluconobacter cerinus	Cerilipin (Ornithin-taurin lipid)
	Agrobacterium tumefaciens	Lysine-containing lipid
	Corynebacterium sp.	Various lipoproteins
	Endomycopsis lipolytica	Various lipoproteins
	Bacillus subtilis	Surfactin, subtilisin, subsporin
	Bacillus licheniformis	Various cyclic lipopeptides
	Bacillus spp	Lipopeptide antibiotics
	Mycobacterium sp.	Fatty acylated nonapeptide
	Nocardia asteroides	Fatty acylated heptapeptide
	Corynebacterium lepus	Lipopeptide
	Corynebacterium sp.	Protein-lipid-carbohydrate complex
	Streptomyces sp.	Amphomycin, Siolipin
	Candida petrophilum	Lipopeptide, protein emulsifier
	Torulopsis petrophilum	Protein emulsifier
	Pseudomonas aeruginosa	« Protein-like activator for n-alkane oxidation »
	Candida lipolytica	Liposan
Polysaccharide lipid complexes	*Candida tropicalis*	Polysaccharide-fatty acid complex
	Acinetobacter calcoaceticus	Emulsan
	Pseudomonas fluorescens	Protein-carbohydrate complex
	Streptococcus sanguis	Lipoteichoic acid

The world market for detergents is increasing very fast and the global consumption of surfactants by the year 2000 will reach an estimated 10 billion kg [18]. Biosurfactants have the potential to replace chemically synthesized tensides in many areas of applications. They are less toxic than synthetic tensides, they are biodegradable, and can be produced on renewable substrates. Novel compounds may be more effective for specific purposes [19]. In addition, the chemical structure and physical properties of biosurfactants can be modified by either genetic, biological, or chemical manipulations allowing one to tailor biosurfactants to specific needs [3]. The keys to a successful industrial production of microbial surfactants are cheap substrates and processes, high biosurfactant yields, easy product recovery and highly active biosurfactants with specific properties [11]. A recent study on the economics of biosurfactants indicates that they are still more expensive than their synthetic counterparts but have a lower critical micelle concentration, thus being more efficient [20].

2 *Pseudomonas aeruginosa* Rhamnolipids

2.1 Composition of Different Rhamnolipids

The rhamnose-containing glycolipid biosurfactants produced by *P. aeruginosa* were first described in 1949 [21]. L-Rhamnosyl-L-rhamnosyl-β-hydroxyde-canoyl-β-hydroxydecanoate and L-rhamnosyl-β-hydroxydecanoyl-β-hydroxy-decanoate, referred to as rhamnolipids 1 and 2, respectively, are the principal glycolipids produced in liquid cultures. Rhamnolipid 1 was also found in the culture supernatants of the *P. aeruginosa* S_7B_1 strain grown on *n*-hexadecane, *n*-paraffin mixtures, or glucose at 30 °C for 6 days [22]. Rhamnolipid 2 was observed during cultivation of *P. aeruginosa* KY4025 on 10% *n*-paraffin at 30 °C for 55 h [23]. Rhamnolipids 3 and 4, containing only one β-hydroxydecanoyl moiety, were detected in culture supernatants of resting cells, however, these types may represent degradation products derived from hydrolysis of rham-nolipids 1 and 2 [24]. Methyl ester derivatives of rhamnolipids 1 and 2 were purified from *P. aeruginosa* strain 158 [25]. Additional types of rhamnolipids harbouring alternative fatty acid chains have been purified from cultures of a clinical isolate of *P. aeruginosa* [26]. The fatty acid homologues present in these rhamnolipids were identified by fast atom bombardment and electron impact mass spectrometry as β-hydroxyoctanoyl-β-hydroxydecanoate, β-hy-droxydecanoyl-β-hydroxydodecanoate, and β-hydroxydecanoyl-β-hydroxy-dodec-5-enoate. The predominant types of rhamnolipids appear to be strain-specific and seem to depend also, to a limited extent, on the environmental and cultivation conditions, especially on the medium composition. The glycosyl moiety has always been found to be built up of rhamnose units, whereas differences in the rhamnolipid structures concern the fatty acid residues. Alternative fatty acids may be formed as intermediates during cultivation on certain hydrocarbon sources. However, our observations indicate that the fatty acid present in the *P. aeruginosa* PG201 rhamnolipids is always β-hydroxy-decanoate, independent of the chain length of the alkanes which are used as the carbon source.

2.2 Biosynthesis of Rhamnolipids

The biosynthesis of the *P. aeruginosa* rhamnolipids was initially studied in vivo by using various radioactive precursors such as [^{14}C]-acetate and [^{14}C]-glycerol [27–29]. A putative biosynthetic pathway has been proposed by Burger et al. [30, 31] and is shown in Fig. 1. In this pathway, the synthesis of rhamnolipids proceeds by sequential glycosyl transfer reactions, each catalyzed by a specific rhamnosyltransferase with thymidine-diphospho-rhamnose (TDP-rhamnose) acting as a rhamnosyl donor and β-hydroxydecanoyl-β-hydroxy-decanoate or L-rhamnosyl-β-hydroxydecanoyl-β-hydroxydecanoate acting as the acceptor. The donor substrate for rhamnolipid biosynthesis, TDP-rhamnose, occurs in many Gram-negative species since rhamnose is often incorporated into the lipopolysaccharide side chains. TDP-glucose is formed through several enzymatic steps from glucose and is converted to TDP-rhamnose by dehydration, epimerization and reduction of the glycosyl moiety [32–34].

The acceptor substrate precursor, β-hydroxydecanoic acid, can be formed by two possible routes [35]. First, it can arise as an intermediate of fatty acid degradation via the β-oxidation cycle. This route is assumed to be the predominant pathway during growth on *n*-alkanes. Second, β-hydroxydecanoic acid occurs as an intermediate during de novo fatty acid biosynthesis. A dimer consisting of two β-hydroxydecanoic acid molecules is formed by condensation, however, the exact mechanism of this esterification is not understood yet.

2.3 Environmental Regulation of Rhamnolipid Biosynthesis

The production of *P. aeruginosa* rhamnolipids depends on several nutritional and environmental factors [36]. Maximal rhamnolipid synthesis is found during the late-exponential and stationary phases of growth under conditions of nitrogen limitation, and a direct relationship between increased glutamine synthetase activity and enhanced biosurfactant production has been demonstrated [37]. Rhamnolipid synthesis is favored under non-limiting concentrations of phosphate [38]. Among the mineral salts in the culture medium, iron has the most profound influence on rhamnolipid synthesis. A three-fold increase in rhamnolipid production without significant changes in biomass yield has been reported after a shift to iron-limiting conditions [39]. Iron limitation has been recognized as a quite general stimulus for the production of various secondary metabolites and virulence factors of a large number of microorganisms. The

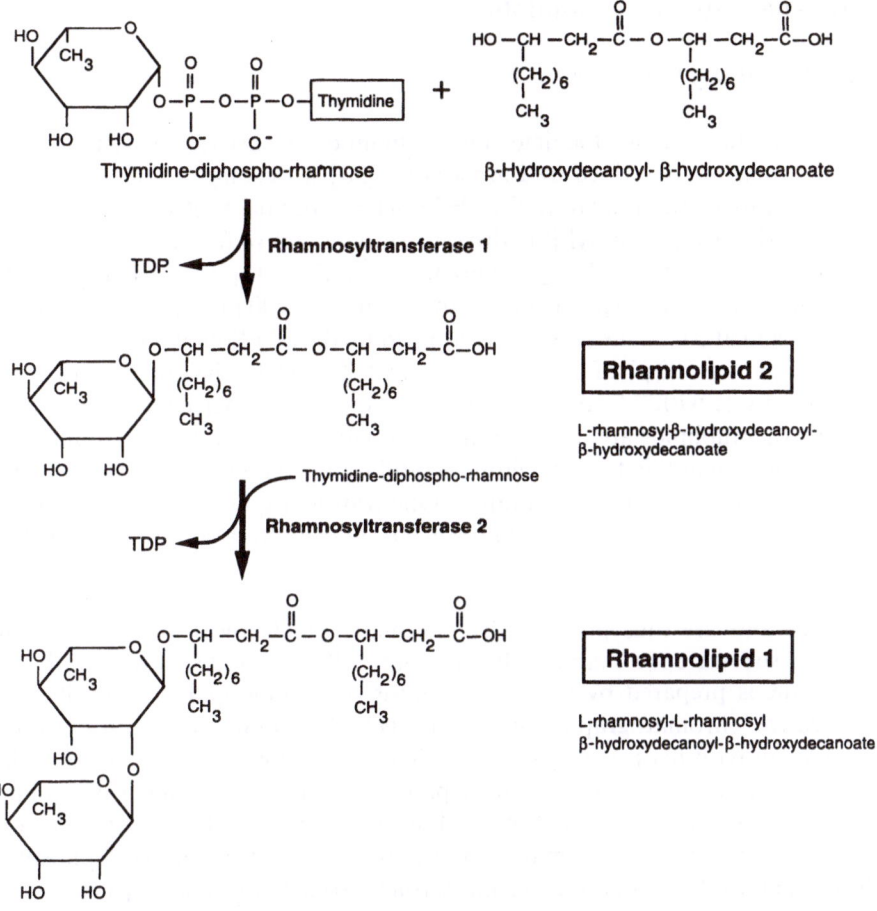

Fig. 1. Biosynthesis of *P. aeruginosa* rhamnolipids. Thymidine-diphosphorhamnose (TDP-rhamnose) acts as a donor substrate in two sequential rhamnosyltransferase reactions. β-hydroxydecanoyl-β-hydroxydecanoate and L-rhamnosyl-β-hydroxydecanoyl-β-hydroxydecanoate are the acceptor substrates

optimal pH and temperature for rhamnolipid formation were determined in continuous culture [39]. Maximal rhamnolipid synthesis was found at pH 6.0–6.5, whereas pH values above 7.0 led to a sharp decrease in the production of surface-active compounds. The temperature optimum was 31–34 °C. Below 30 °C or above 37 °C, the rhamnolipid yields were significantly reduced. The knowledge of the chemical and physical parameters which are most critical for rhamnolipid production allowed the design of an optimized medium and process which we further modified as described in Sects. 6 and 7 of this review. Furthermore, the elucidation of the biochemistry and genetics of rhamnolipid biosynthesis, as described in the following chapters, allows us to understand its complex regulation at the molecular level.

3 Biochemistry of Rhamnolipid Synthesis

3.1 Rhamnosyltransferase Assay

The enzymatic assay for the detection of rhamnosyltransferase activity is based
on the procedure described by Burger et al. [30] with some modifications. The
assay mixture contains 50 mM Tris $\cdot$ HCl pH 7.5, 10 mM $MgCl_2$, 1 mM EDTA,
0.2 mM β-hydroxydecanoyl-β-hydroxydecanoate and/or 0.2 mM L-rhamnosyl-
β-hydroxydecanoyl-β-hydroxydecanoate, 0.2 mM [^{14}C]-TDP-rhamnose (500
cpm nmol^{-1}), and 1–50 µg of enzyme in a volume of 200 µl. After incubation for
up to 60 min at 37 °C, the reaction is stopped by the addition of 300 µg of carrier
rhamnolipid in 100 µl of H_2O and 20 µl of 2 N HCl. The rhamnolipids are
extracted twice with 0.5 ml of diethylether, the pooled ether fractions are dried
and the rhamnolipids are dissolved in 20–100 µl of methanol. The products are
subsequently quantified using a liquid scintillation counter or separated by TLC
analysis followed by autoradiography. One transferase unit corresponds to the
incorporation of 1 nmol of rhamnose from TDP-rhamnose into rhamnolipid
per hour at 37 °C.

The substrates for the rhamnosyltransferase assay are not commercially
available and are thus prepared by enzymatic or chemical means [40]. The
acceptor substrate for rhamnosyltransferase 1, β-hydroxydecanoyl-β-hydroxydecanoate, is prepared by partial hydrolysis of rhamnolipids and purified by
preparative chromatography on a Silicagel 60 column. L-rhamnosyl-β-hydroxydecanoyl-β-hydroxydecanoate, which acts as the acceptor substrate for
rhamnosyltransferase 2, is purified by preparative TLC of a crude rhamnolipid
mixture. [^{14}C]-labeled thymidine-diphospho-rhamnose (TDPR) can be obtained by preparative enzymatic conversion of [^{14}C]-thymidine-diphospho-
glucose (TDPG) using crude enzyme extracts from *P. aeruginosa* [30].

3.2 Rhamnosyltransferase Activity in P. aeruginosa Extracts

The typical time course of growth, rhamnosyltransferase activity and rhamnolipid production during a batch cultivation of *P. aeruginosa*. PG201 is shown
in Fig. 2. A phosphate-buffered modification of medium M3 [39] with 2% of
glycerol as the carbon source, referred to as GS-medium, is used. Maximal
rhamnosyltransferase activity is detected after the cells enter the stationary
phase of growth. Three different *P. aeruginosa* wild-type strains, PG201 [39],
ATCC7700 [30] and PAO1 [41] show a very similar behavior regarding
rhamnolipid formation and transferase activities.

3.3 Characterization of Rhamnosyltransferases

Rhamnosyltransferase 1 activity is found in the soluble enzyme fraction as well
as in the membrane and particulate fractions, suggesting that rhamnosyltran-

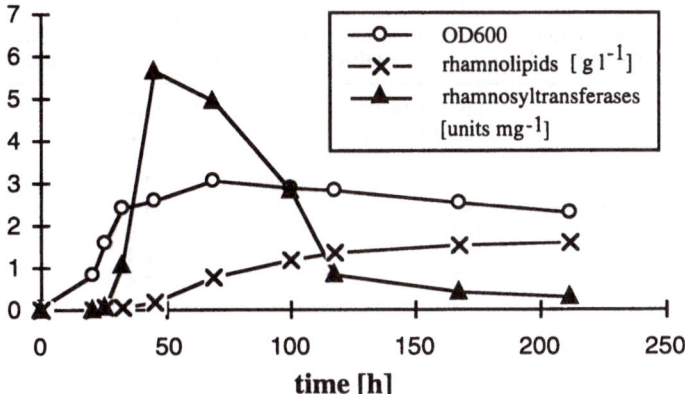

Fig. 2. Growth, rhamnolipid concentration in the supernatant and rhamnosyltransferase activity during cultivation of *P. aeruginosa* PG201 in nitrogen limiting minimal medium containing glycerol as the carbon source.

sferase 1 is membrane-associated and is partially liberated during cell fractionation procedures. Separate protocols for the purification of the soluble and the membrane-bound rhamnosyltransferase 1 have been developed [40]. The soluble enzyme can be purified by precipitation with 40% saturated ammonium sulfate and subsequent chromatography on hydroxyapatite, DEAE-cellulose and phenyl-Sepharose (Table 2). However, a high amount of the enzymatic activity is lost during these sequential purification steps since rhamnosyltransferase 1 exists as a complex built from two separate subunits which may separate and become inactive during the protein purification procedures.

A considerable amount of the soluble rhamnosyltransferase 1 activity is present in high-molecular-weight aggregates which can be separated from the bulk proteins in the crude soluble fraction by gel filtration on Sepharose CL-6B (Fig. 3). Rhamnosyltransferase 1 activity is found primarily in the first protein-containing fractions eluted from the column, consisting of large complexes with a molecular weight of at least 2×10^6. These rhamnosyltransferase complexes are absolutely stable when subjected to concentration and re-chromatography on the same column. The enrichment of rhamnosyltransferase by gel filtration is 16-fold, with a yield of 65%. The complexes contain a considerable amount of LPS, which cannot be removed from the protein without a severe loss of enzymatic activity. The observation that rhamnosyltransferase is associated with LPS gives further evidence that at least some of the transferase activity is localized in the cell envelope in vivo.

The membrane-associated rhamnosyltransferase 1 activity can be purified by extraction of the cell membranes with 1 M KCl [40]. This procedure is known to release selectively peripheral membrane proteins [42]. The KCl-extracted proteins are further purified by hydrophobic interaction chromatography [40].

Table 2. Purification of soluble rhamnosyltransferase 1. The soluble enzymes after ultracentrifugation were precipitated with ammonium sulfate, desalted by gel filtration on Sephadex G-50 and further purified by chromatography on hydroxyapatite, DEAE-cellulose and phenyl-Sepharose

fraction	Spec. activity [units mg^{-1}]	Yield %	Enrichment factor
Soluble enzyme	28	100	1
Ammonium sulfate precipitate, desalted	49.5	95	1.8
Hydroxyapatite pool	163	22	5.8
DEAE-cellulose pool	235	13	8.4
Phenyl-Sepharose pool	432	2	15.4

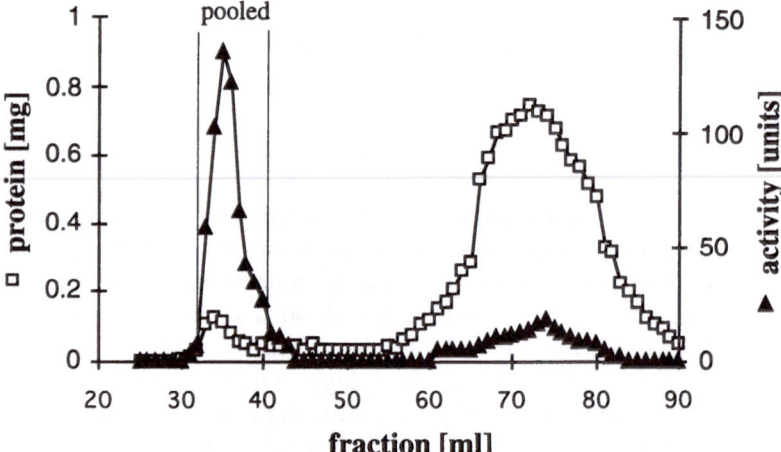

Fig. 3. Purification of high-molecular-weight rhamnosyltransferase complexes by gel filtration of soluble proteins of Sepharose CL-6B. Rhamnosyltransferase 1 activity was enriched 16-fold in the fractions eluted just after the void volume of the column

SDS-PAGE analysis of such enriched rhamnosyltransferase 1 reveals one major protein of roughly 47 kDa. It will be shown later that the catalytic subunit of the rhamnosyltransferase complex is indeed a protein of 47 kDa.

Rhamnosyltransferase 2 activity is found mainly in the particulate fraction after cell disintegration and is therefore inaccessible for purification. The small portion of rhamnosyltransferase 2 present in the soluble enzyme fraction can be partially purified by precipitation with 0.33% protamine sulfate. Under these conditions, rhamnosyltransferase 1 activity remains in the supernatant. The recovery of active rhamnosyltransferase 2 from the protamine precipitate is achieved by solubilization with 0.1% Triton X-100/0.5 M NaCl. Anionic detergents such as deoxycholate or SDS, however, lead to a loss of enzymatic activity.

4 Genetics of Rhamnolipid Synthesis

The overexpression of genes involved in the biosynthesis of surfactants and the control of these genes is an important goal to be achieved towards an economical production of biosurfactants on an industrial scale. Only little information is available about the genetics of biosurfactant synthesis. Phenotypes of mutant strains which are positively or negatively affected in the production of surfactants have been described for a variety of bacteria, including the rhamnolipid forming *P. aeruginosa*, the surfactin producing *Bacillus subtilis*, the emulsan-producing *Acinetobacter calcoaceticus*, and the serraphobin and serratamolide producing *Serratia marcescens*, as recently reviewed by Reiser et al. [43]. *P. aeruginosa* is the genetically best studied Gram-negative surfactant producer. Among Gram-positive species, similarly detailed genetic studies focusing beyond the description of mutant strains are available for *Bacillus subtilis* [44–46].

4.1 Mutants Affected in Rhamnolipid Synthesis

Our strategy towards the identification of the genes encoding the rhamnolipid biosynthetic pathway was based on the isolation of rhamnolipid-nonproducing mutants followed by genetic complementation of such strains using wild-type genes. A pool of Tn5-GM induced mutants of *P. aeruginosa* PG201 harboring a single randomly inserted transposon in the genome was generated as described by Koch et al. [47]. A total of 70 000 mutants were screened for their ability to produce rhamnolipids by using the methylene blue/CTAB agar plate assay described by Siegmund and Wagner [48]. Putative rhamnolipid-deficient mutants were then tested for their capacity to synthesize the biosurfactants during cultivation in liquid GS-medium containing either 2% glycerol or 1% hexadecane as the C-source. The physiological and biochemical analysis of some rhamnolipid-deficient mutants is shown in Table 3, together with the data of the

Table 3. Physiological and biochemical characterization of *P. aeruginosa* wild-type strains (PG201, ATCC7700, PAO1) and of Tn5-GM induced rhamnolipid-deficient mutants (UO287, UO299, UO391, 65E12). The rhamnolipid concentrations in the culture supernatant and the rhamnosyltransferase activities in crude cell extracts were determined after 6 d of cultivation at 37 °C in GS-medium containing 2% glycerol

Strain	Growth [OD600]	Rhamnolipids [$g \, l^{-1}$]	Rhamnosyltransferase [units mg^{-1}]	Utilization of hexadecane
PG201	3.13	1.37	45	+ +
ATCC7700	3.05	1.21	48	+ +
PAO1	5.14	1.64	24	+ +
UO287	3.60	< 0.005	< 0.5	+
UO299	3.29	< 0.005	< 0.5	+
UO391	3.33	< 0.005	< 0.5	+
65E12	1.29	< 0.005	< 0.5	−

previously described 65E12 mutant strain [47], the PG201 parent strain [39] and two other *P. aeruginosa* wild-type strains, PAO1 [41] and ATCC7700 [30], respectively. Three strains, UO287, UO299, and UO391, completely lacked rhamnolipids and rhamnosyltransferase under all tested conditions. However, these strains were still capable of using hexadecane as the sole carbon and energy source, but their growth rate in hexadecane containing media was significantly reduced and the initial lag phase was much longer than *P. aeruginosa* wild-type strains. These findings suggest that rhamnolipids are not essential for growth on hydrocarbons but facilitate the utilization of these water-insoluble compounds by efficient solubilization.

4.2 Genes Involved in Rhamnolipid Biosynthesis

The isolation of well-defined rhamnolipid-negative mutant strains as described above facilitated the identification of the genes involved in rhamnolipid biosynthesis. Tn5 insertions generally occur only at a single site in the genome and the DNA sequence of the transposon Tn5 including that of the distal IS50 elements is known. Therefore, DNA fragments containing the transposon flanking region were easily subcloned and the DNA sequence at the site of insertion was determined by using a Tn5 specific sequencing primer. DNA sequence homologies with any known genes in the data library were not detected,

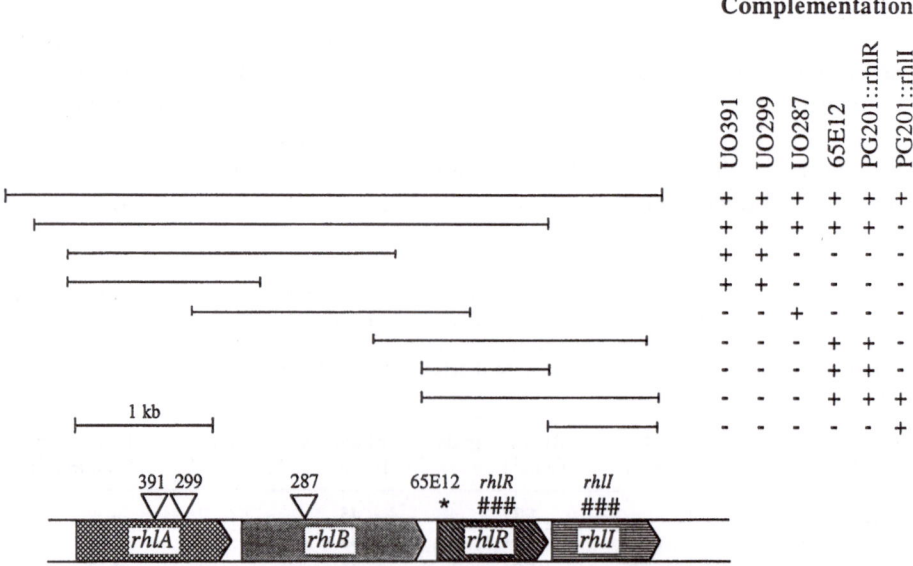

Fig. 4. Gene organization in the *rhl* gene cluster relevant for rhamnolipid production in *P. aeruginosa*. The four open reading frams are designated *rhlA*, *rhlB*, *rhlR*, and *rhlI*, respectively. The sites of mutation in the rhamnolipid-deficient strains and the genetic complementation analysis of these mutants by various subfragments of the *rhl* cluster are indicated

therefore, the newly identified genes were designated as *rhl* genes. The corresponding *rhl* wild-type genes were isolated by using the transposon flanking DNA-regions of the rhamnolipid-deficient mutants as hybridization probes to screen a cosmid gene library that had previously been constructed [49]. Three cosmids were isolated which were capable of restoring rhamnolipid biosynthesis in all of the rhamnolipid mutant strains, indicating that the mutations in these different strains affected the same gene cluster. Further subcloning and sequence analysis of the complementing cosmids revealed a cluster of four open reading frames (ORFs) within the *rhl* region all of which are relevant for rhamnolipid production in *P. aeruginosa*. The site of transposon insertion in the mutant strains UO299 and UO391 affected ORF1, whereas the UO287 strain carried a transposon in ORF2. The 65E12 mutant strain harbours a single base pair deletion in ORF3. The construction of additional mutants by insertional mutagenesis of ORF3 and ORF4 proved that both of these putative genes are absolutely required for rhamnolipid biosynthesis. Various subfragments of the *rhl* gene cluster were tested for their capacity to complement the genetic defects in the rhamnolipid-deficient mutants. Four complementation groups were found, which is consistent with the identification of four putative genes in the *rhl* cluster. The gene organization in the *rhl* region, with ORF1 to ORF4 termed *rhlABRI*, is depicted in Fig. 4, together with the sites of mutation and the complementation analysis of the rhamnolipid-deficient mutants.

4.3 Properties of the Rhamnolipid Genes and Their Products

The codon usage in the *rhlABRI* genes agrees very well with the typical codon preferences reported for *P. aeruginosa* genes [50]. The functions of the *rhl* gene products were elucidated by combined analyses involving heterologous expression of the genes and homology studies with proteins of a known function. The characterization of the *rhl* genes and their products is shown in Table 4. The *rhlAB* genes are organized as an operon and encode the two subunits constituting a functional rhamnosyltransferase [51]. The 32.5 kDa RhlA protein harbors an *N*-terminal amino acid sequence that shares similarities with typical signal

Table 4. Properties of the *rhlABRI* genes and their products

| Gene | DNA Sequence data | | | Corresponding protein encoded | | |
	%(G + C)	size [nt]	promoter	[kDa]	pI	putative function
rhlA	65.8	887	σ^{54}	32.5	7.4	rhamnosyltransferase subunit
rhlB	67.9	1280	σ^{54}	47	8.4	rhamnosyltransferase subunit
rhlR	61.7	726	σ^{70}	26.5	7.0	transcriptional activator of the LuxR family
rhlI	64.8	603	n.d.	21	n.d.	autoinducer synthetase of the LuxI family

peptides found in Gram-negative bacteria, suggesting that this subunit is exported to the periplasm. The RhlB protein is the catalytic subunit of the rhamnosyltransferase complex. It has a molecular weight of 47 kDa, as deduced from the *rhlB* nucleotide sequence. A hydropathy plot of the RhlB protein revealed the presence of at least two membrane-spanning domains, one including the first 20 amino acids at the *N*-terminus of the RhlB protein, and the second being located in the *C*-terminal part of the protein, suggesting that the RhlB protein is membrane anchored via its *N*-terminus and that it crosses the membrane, thus exposing its domains on both sides of the membrane. These findings are consistent with previous results obtained during the purification of the rhamnosyltransferase, where a 47 kDa membrane protein with rhamnosyltransferase activity had been purified.

The RhlR and RhlI proteins control the expression of the *rhlAB* rhamnolipid biosynthetic genes [52, 53]. The RhlR protein is a transcriptional activator of the LuxR family. The *C*-terminal part of these proteins contains a helix-turn-helix DNA binding motif, which appears to be conserved in transcriptional activators [54]. The activity of the RhlR regulator is mediated by a small diffusible molecule termed autoinducer which is produced by the RhlI protein. The RhlR-RhlI protein pair has a high degree of homology to the LasR-LasI proteins controlling elastase production in *P. aeruginosa* and the LuxR-LuxI proteins regulating bioluminescence in *Vibrio fischeri*. These so-called quorum-sensing regulatory systems act in a cell-density and starvation dependent fashion, allowing the expression of their target genes under certain physiological and environmental conditions [55].

4.4 Regulation of Rhamnolipid Gene Expression

The production of extracellular rhamnolipids by *P. aeruginosa* occurs under conditions of nitrogen and iron limitation during the late-exponential and stationary phases of growth [36, 37]. These findings suggested early on that the genes encoding the rhamnolipid biosynthetic pathway are strictly regulated at the transcriptional level. The expression of the *rhl* genes was studied by using translational fusions of the *rhl* promoter regions to the *lacZ* reporter gene, thereby allowing the detection of *rhl* gene expression by measuring β-galactosidase activity (Table 5). The *rhlA* but not the *rhlB* promoter is active, suggesting a bicistronic transcript covering the coding regions of both the *rhlA* and *rhlB* genes. The expression of the *rhlAB* operon is enhanced 20-fold during the stationary phase of growth under conditions of nitrogen limitation. This *rhlA* promoter activation was absent in mutant strains defective in either the *rhlR* or *rhlI* locus, indicating that a functional RhlR-RhlI regulatory system is necessary for the transcriptional activation of the *rhlAB* operon [53]. The *rhlA* promoter is recognized by σ^{54}, which is an alternative sigma factor of RNA polymerase [56]. Strains lacking the *rpoN* gene encoding this sigma factor are incapable of producing rhamnolipids. The regulatory DNA sequence upstream

Table 5. Activities of the *rhl* gene promoters. The expression of the *rhl* genes in *P. aeruginosa* was determined by using promoter fusions to the *lacZ* gene. The cells were cultivated in LB complete medium or in minimal medium with limiting amounts of nitrogen and iron

Promoter fusion	β-Galactosidase activity in permeabilized cells [Miller units]		
	LB Medium	Minimal medium with N- and Fe-limitation	
		Exponential phase	Stationary phase
None	< 10	< 10	< 10
rhlA' :: *lacZ*	180	680	13600
rhlB' :: *lacZ*	< 10	< 10	30
rhlAB' :: *lacZ*	< 10	250	1560
rhlR' :: *lacZ*	150	50	160

of the *rhlA* promoter contains two inverted repeats both of which are ultimately required for *rhlA* gene expression. They carry the common GTTC-N_{13}-GAAC sequence motif defining putative binding sites for the RhlR regulator [51]. The *rhlR* regulatory gene is constitutively transcribed from a σ^{70} promoter at low levels. The RhlR protein becomes a functional transcriptional activator after post-translational modification through binding to a small diffusible signalling molecule, termed autoinducer. Several similar autoinducer-dependent regulatory systems have been described in various bacteria, including *Vibrio* species, *P. aeruginosa*, *Erwinia carotovora*, and *Agrobacterium tumefaciens* [55]. In all of these systems, *N*-acyl homoserine lactones act as the signalling autoinducer molecule. All of the signalling molecules known so far have a common homoserine lactone ring structure, whereas the length of the acyl chain varies between the different systems. During growth of the bacteria, the autoinducer molecules slowly accumulate and are present in a high concentration during the late-exponential phase of growth when cell densities become high. Above a certain threshold concentration, the signalling molecules bind to a cognate regulatory protein which then activates its target genes. Autoinducer-dependent regulatory systems provide a genuine mechanism for cell-to-cell communication and allow the expression of certain genes only under conditions of high cell densities or of starvation. Such genes encode extracellular compounds including degradative enzymes, virulence factors, and enzymes required for the production of biosurfactants, which have more advantageous effects in a dense bacterial population. The autoinducers involved in the regulation of rhamnolipid production are generated by the RhlI autoinducer sythetase. The *rhlI* gene is itself positively regulated by the RhlR protein [53]. A mutant strain with a disrupted *rhlI* gene does not produce detectable levels of rhamnolipids, but can be stimulated by the addition of synthetic *N*-acyl homoserine lactones or of small amounts of spent wild-type culture medium. These stimulation experiments indicate that *N*-butyryl homoserine lactone and *N*-(3-oxohexanoyl) homoserine lactone are the rhamnolipid autoinducer, whereas a related compound, *N*-(3-oxododecanoyl)-homoserine lactone, which occurs in *P. aeruginosa* as well, was

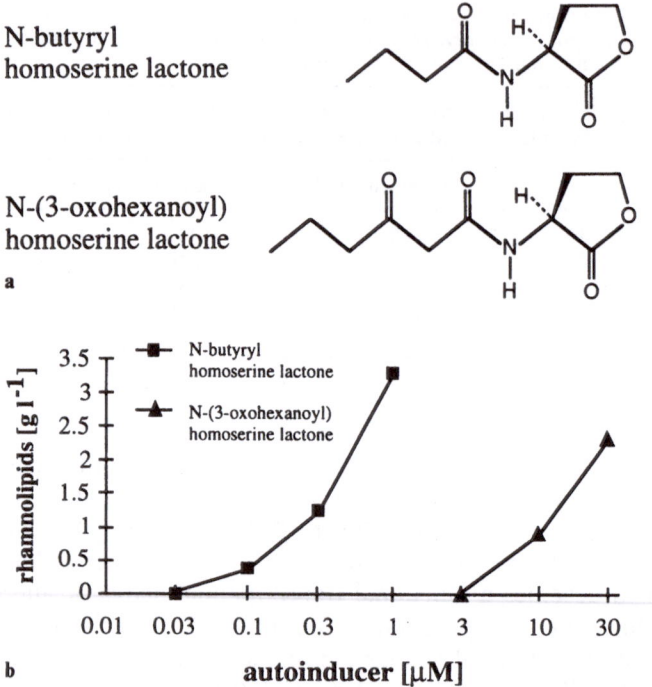

N-butyryl homoserine lactone

N-(3-oxohexanoyl) homoserine lactone

a

b   **autoinducer [μM]**

Fig. 5a, b. Autoinducer-dependent stimulation of rhamnolipid production in *P. aeruginosa*. **a** Chemical structures of *N*-butyryl homoserine lactone and *N*-(3-oxohexanoyl) homoserine lactone. **b** Stimulation of rhamnolipid synthesis in a *P. aeruginosa* mutant strain affected in its *rhlI* autoinducer synthetase gene. Increasing amounts of synthetic *N*-butryl homoserine lactone or *N*(3-oxohexanoyl) homoserine lactone were added during the early exponential growth phase (OD600 = 0.15), and rhamnolipid concentrations were determined after 5 d of cultivation in GS minimal medium

less efficient (Fig. 5). The regulation of rhamnolipid production in *P. aeruginosa*, as presented in a tentative model shown in Fig. 6, provides therefore an excellent example for the complex mechanisms controlling microbial surfactant synthesis, which is a high energy and carbon requiring process.

4.5 Production of Rhamnolipids in Recombinant Hosts

The expression of the rhamnosyltransferase genes in a variety of heterologous hosts has recently been studied [57]. In these experiments, the intact *rhlABRI* gene cluster was cloned into the broad-host-range plasmid pJRD215 [58] (Fig. 7A). In this construct (pUO101) the rhamnosyltransferase *rhlAB* operon is expressed from its own promoter which is controlled by the *rhlR* and *rhlI* gene products as described in the previous section. In *P. aeruginosa* PG201, the presence of additional plasmid-borne *rhl* gene copies leads to a roughly 1.5-fold

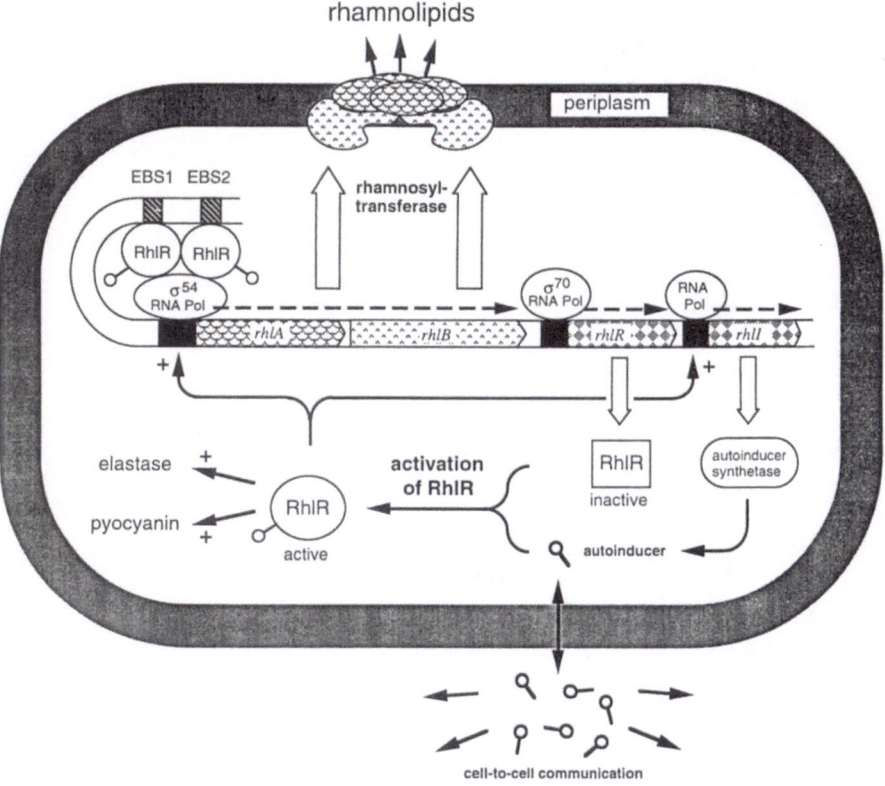

Fig. 6. Model for the regulation of rhamnolipid production in *P. aeruginosa*. The transcriptional activation of the *rhlAB* rhamnosyltransferase genes involves the quorum-sensing RhlR-RhlI regulatory system

increase in rhamnolipid formation and rhamnosyltransferase activity (Table 6). Among the heterologous hosts containing the *rhl* gene cluster, *P. fluorescens* produces rhamnolipids, but its concentration in the culture supernatant is only 10% of that obtained with the *P. aeruginosa* strain, indicating that additional factors of *P. aeruginosa* may be involved in the regulation of rhamnolipid production and that these factors may be missing in heterologous hosts. Alternatively, a plasmid was constructed which contained the promoterless *rhlAB* genes downstream of the *tac* promoter in the broad-host-range expression vector pVLT35 [59], thus allowing the inducible expression of the *rhlAB* genes by IPTG added to the growth medium (Fig. 7B). Using this construct (pUO98), the highest levels of rhamnosyltransferase (45 units ml^{-1}) and of rhamnolipids (0.6 g l^{-1}) are obtained in recombinant *P. putida* KT2442 grown in LB medium supplemented with 1% glucose. Product formation occurs within one hour after induction and the volumetric productivity of biosurfactants reaches maximal levels of 24 mg l^{-1} h^{-1} during the exponential growth phase of this recom-

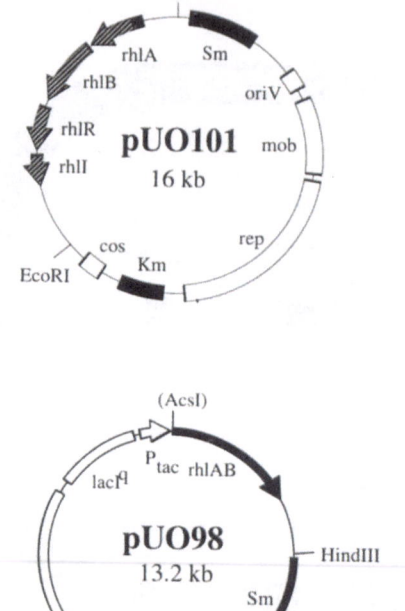

a

b

Ptac > AATTTTTGGGAGGTGTGAAATG...

Fig. 7a, b. Plasmids used for the production of rhamnolipids in heterologous hosts. **a** Plasmid pUO101 harbours a 5.8 kb DNA fragment containing the *rhlABRI* gene cluster that was cloned into the broad-host range vector pJRD215. **b** In pUO98, a 3.1 kb DNA fragment containing the promoterless *rhlAB* rhamnosyltransferase genes was cloned downstream of the *tac* promoter in pVLT35

binant strain. As expected, the induced *P. putida* strain forms only rhamnolipid 2 which is composed of one rhamnose unit per biosurfactant molecule and which is the product of the rhamnosyltransferase 1 reaction. *P. putida* KT2442 is known to produce high levels of poly(3-hydroxyalkanoates), even when grown on glucose as the carbon source, whereas other fluorescent Pseudomonads are generally less efficient in this respect [60]. The fatty acid metabolism may be disregulated in this strain, with accumulated fatty acids serving as precursors for PHA and rhamnolipid synthesis, thereby making this strain very useful as a host for heterologous rhamnolipid production. Rhamnolipids can be produced in heterologous strains provided that the rhamnosyltransferase genes are efficiently expressed. Further medium and bioprocess optimization may lead to substantially higher biosurfactant productivities using strains of the G.R.A.S status which can be applied industrially.

Table 6. Rhamnosyltransferase activity and rhamnolipid formation in recombinant Pseudomonads. Plasmid pUO101 contains the *rhlABRI* gene cluster. Cells harbouring pUO101 were cultivated for 7 d in GS minimal medium with 2% glycerol. Plasmid pUO98 contains the *rhlAB* operon under the control of the inducible *tac* promoter. Cells harbouring UO98 were induced during the exponential growth and further cultivated for 19 h in LB medium supplemented with 1% glucose

Strain/Plasmid	Rhamnosyltransferase [units ml^{-1}]	Rhamnolipids [g l^{-1}]
P. aeruginosa		
PG201/pUO101	24 ± 4	2.2 ± 0.4
PG201/pJRD215	18 ± 4	1.6 ± 0.4
P. fluorescens		
ATCC15453/pUO101	4.0 ± 0.5	0.25 ± 0.03
ATCC15453/pJRD215	1.2 ± 0.5	0.06 ± 0.03
P. aeruginosa		
PG201/pUO98	15 ± 3	0.15 ± 0.05
P. putida		
KT2442/pUO98	45 ± 5	0.60 ± 0.15

5 Prospects and Limits for the Industrial Production of Rhamnolipids

Towards the development of a rhamnolipid production process on an industrial scale, the relevant process parameters have to be identified and should be optimized from a commercial point of view. If it is intended to sell the product, the acceptance of the customers must receive the appropriate attention. Profound results concerning the research and development, as they are now available in the case of rhamnolipids, will be most helpful.

The aim of the first attempts of rhamnolipid production in shake flasks was to obtain some material for chemical and physical investigations [21, 27]. The development of biochemical engineering concepts with the specific purpose of producing rhamnolipids started about 15 years ago and included batch [61], fed-batch [62], and continuous processes without [39, 63] or with [64, 65] cell recycling. The use of carrier-fixed resting cells was also tested [66]. Production problems during the first processes occurred due to several reasons:

- Excessive foam formation, caused by bubble aeration
- Poor product yield due to non-optimized media
- High volumetric productivities due to low growth rates.

Meanwhile, these production problems are solved. Several possibilities to destroy the foam exist:

- Foam-breakers (mechanical)
- Antifoam agents (chemical)
- Integration of the foam formation into the chemical engineering concept.

Mechanical tools were not sufficient to control the foam formation at high rhamnolipid concentration. Stable working states in a continuous system could not be obtained when the surfactant concentration exceeded $2.25 \, gl^{-1}$ [63]. Under such conditions, the stability of the rhamnolipid foam was further increased by proteins and microbial cells. The alternative solution to the foam problem, the use of anifoam agents, is not economical, since the costs for antifoam agents increase the costs of surfactant production. The first integrated closed production process controlling the foam problem was developed by Gruber [64]. In this system, the carbon dioxide was removed by using non-porous solution-diffusion membranes and the foam arising was integrated in the production process, resulting in a better exploitation of the reactor volume. Ways to enhance the product yield include:

- Strain improvement
- Medium development
- Process optimization

A first step towards the strain improvement was the isolation of overproducing *P. aeruginosa* strains [38, 49]. The use of genetic engineering methods for strain improvement was described in Sect. 4 above. Chemical and physical parameters that influence rhamnolipid synthesis have been systematically investigated in order to enhance the productivity [39]. A substantial increase in rhamnolipid production was observed under conditions of nitrogen limitation or multivalent cation limitation after the exponential growth phase. Possible advantages of using different carbon sources, such as *n*-paraffin [61], corn oil [62, 65], or olive oil [67] have been described in detailed studies. Temperature shifts and pH shifts were also applied to enhance the productivity [10].

To save investment costs and operating results in microbial production processes, high product yields and high volumetric productivities are striven for. The transition from batch to continuous processes and the increased cell densities due to cell recycling systems resulted in a substantially enhanced volumetric productivity. Up to now, the highest volumetric productivities, based on either water-soluble [64] or water-insoluble [65] substrates, were achieved by using a production system where the substrate residence time was separated from the cell residence time by a membrane supported cell recycling system. The relevant steps in the development of this integrated system are shown in the following sections.

6 Medium Design

The composition of the medium depends on the requirements of the production strain as well as on the process conditions. Therefore, a multi-stage approach appears to be the most suitable method for medium design [65]. First of all, the qualitative factors that significantly influence rhamnolipid production have to be identified. These basic experiments are carried out in a continuous culture using the pulse and shift techniques. Since medium optimization and process parameters influence each other reciprocally, this step is a protracted dynamical process. The factors of utmost relevance for rhamnolipid production are presented below.

6.1 Influence of Carbon Source in Biosurfactant Production

Rhamnolipids are formed by *P. aeruginosa* grown on different substrates such as glycerol [21], mannitol, fructose [69], glucose [39, 63, 64], *n*-paraffin [61] and vegetable oils [62, 65, 67, 70]. The product yields, as described in these studies, reach from $Y = 0.1 \, \mathrm{g \, g^{-1}}$ for glucose as the carbon source [64] up to $Y = 0.48 \, \mathrm{g \, g^{-1}}$ for olive-oil containing substrates [67]. However, the strains used in these production experiments were as different as the mineral salt concentrations in the used media. Therefore, we investigated the influence of the carbon source under otherwise identical medium composition. For *P. aeruginosa* PG201 (DSM2659) grown in medium M1 [39], the highest specific rhamnolipid formation was realized by using vegetable oils (Table 7). Among these, corn oil may be the carbon source of choice, since it is cheaper than olive oil.

6.2 Influence of Nitrogen on Biosurfactant Production

P. aeruginosa is capable of utilizing both ammonia and nitrate as nitrogen sources. However, with strain PG201 and corn oil as the carbon source, a higher biosurfactant production is observed when nitrate serves as the sole nitrogen source (Table 8). Since nitrogen is a macronutrient for microbial growth, a sufficient nitrogen supply is essential for biomass formation. On the other hand, a significant enhancement of rhamnolipid synthesis results from nitrogen

Table 7. Rhamnolipid formation by *P. aeruginosa* PG201 (DSM 2659) using different carbon sources. The cells were cultivated for 72 h at 30 °C in shaking flasks in phosphate-buffered basic medium M 1 [8] containing 0.1% $NaNO_3$ as the nitrogen source

C-Source (2%)	Growth	Specific rhamnolipid formation [g g^{-1}]
glucose	+	1.9
galactose	−	−
lactose	−	−
glycerol	+	2.0
n-paraffin	+	2.7
olive oil	+	3.6
corn oil	+	3.7

limitation (Fig. 8). Nitrogen starvation directs the cellular metabolism towards product formation with rhamnolipid being one of the main released products. The optimization of the C-to-N ratio was achieved in continuous culture experiments, indicating that C-to-N ratios from 15 to 23 are optimal for obtaining a high specific productivity [39,65].

6.3 Influence of Iron on Biosurfactant Production

Rhamnolipid formation is partially repressed under conditions of high iron concentrations. In a continuous culture without cell recycling, the rhamnolipid yield is enhanced if the lack of iron causes growth limitation [65]. The optimized C-to-Fe ratio for the highest rhamnolipid yield is 60 000. The effect of a Fe-pulse on a Fe-limited culture is depicted in Fig. 9.

Table 8. Influence of the nitrogen source on the rhamnolipid production by *P. aeruginosa* PG201 (DSM 2659). The cells were cultivated for 72 h at 30 °C in shaking flasks in phosphate-buffered basic medium M 1 [8] containing 2% corn oil and different nitrogen sources

Nitrogen source	Specific rhamno-lipid formation [g g^{-1}]	Substrate utilization [%]	Volumetric productivity [mg l^{-1} h^{-1}]
5.0 g l^{-1} NaNO$_3$	3.67	83	70
4.0 g l^{-1} (NH$_4$)$_2$SO$_4$	1.38	19	19

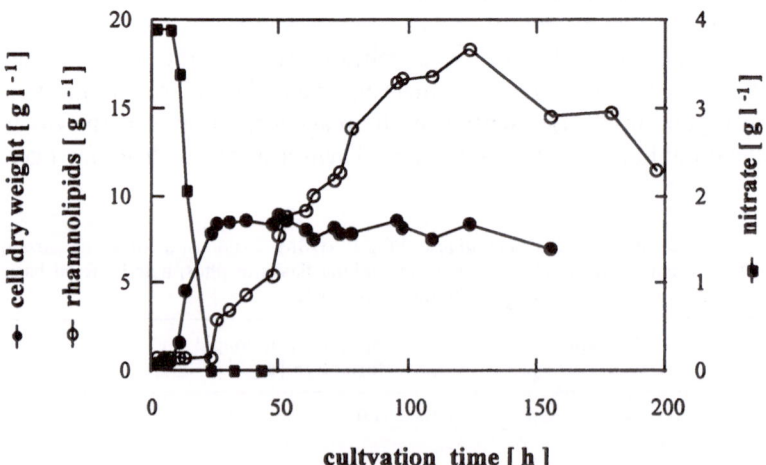

cultvation time [h]

Fig. 8. Growth, rhamnolipid production and nitrate consumption in a batch cultivation of *P. aeruginosa* in a CSTR. Mineral salt medium M2 [66] with a C-to-N ratio of 70 was used. T = 30 °C; pH 6.3; pO$_2$ = 60%. The production phase started when the nitrogen source was used up and extended over a period of 100 h, until the carbon source was consumed

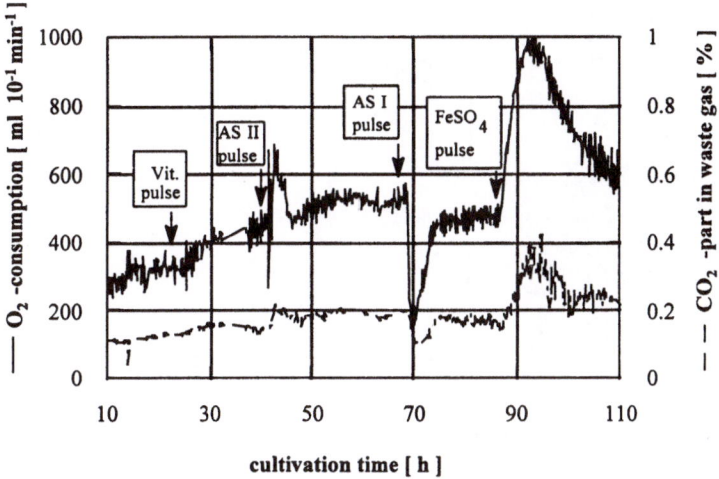

Fig. 9. Pulse experiments in a chemostat. The O_2 consumption and the CO_2 formation are shown. Pulses of vitamins, amino acids, and iron sulphate were applied at the indicated time points. Medium M2 containing 40 g l^{-1} corn oil was used. D = 0.05 h^{-1}, T = 33 °C, pH = 6.3

7 Development of an Integrated Process

The plant for rhamnolipid production in an industrial scale using *P. aeruginosa* PG201 (DSM2659) has been examined and evaluated in detail [64, 65]. The main characteristics of the media were carbon and phosphorus excess as well as nitrogen and iron limitation. The biosurfactant yield was 0.1 g g^{-1} and the productivity 0.5 g l^{-1} h^{-1} when glucose was used as the carbon source. After medium and process optimization and by using corn oil as the carbon source, the product yield increased up to 0.48 g g^{-1} and the productivity was enhanced up to 2 g l^{-1} h^{-1}. The flow scheme of this high performance process is shown in Fig. 10. The plant is based on a continuous stirred tank reactor (CSTR) harbouring an external loop. Two membrane steps are integrated into this loop, one for cell recycling and one for oxygenation and carbon dioxide removal. Corn oil containing mineral salt medium and oxygen is fed into the system. Led away are a cell free permeate stream, a cell containing bleedstream, and the byproduct of metabolism, carbon dioxide. The oxygen partial pressure is controlled by direct entrainment of oxygen into the bioreactor. The microbial growth is controlled by the bleedstream.

7.1 Cell Recycling

The filtration retrains the microorganisms, and, depending on membrane and operation conditions, a certain amount of the product. The increase of the

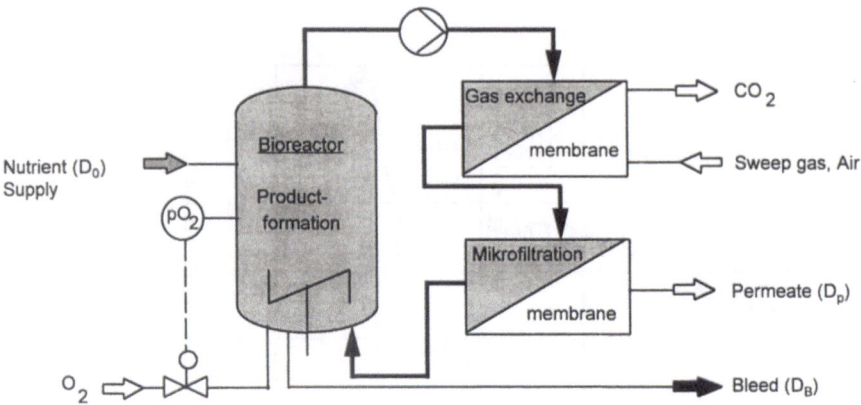

Fig. 10. Flow scheme of a continuous rhamnolipid production process

biomass concentration in a continuous process with cell recycling can be described by the retention R:

$$R = \frac{D_0 - D_B}{D_0}$$

Best results for cell recyling are achieved by using a commercial hollow fiber microfiltration model. The use of a membrane for cell recycling has several advantages:

– No transport limitation occurs, in contrast to systems using carrier-fixed cells [66]
– Cell retention can be varied between 0 and 0.85 even at high biomass concentrations
– Simple handling

The often discussed disadvantages of filtration membranes, such as the decrease of the transmembrane flux and the variation of retention characteristics during operation, are overcome. Using backflush, constant values for flux and retention characteristics are achieved after some hours of operations. A significant retention of biosurfactants by ultra- or microfiltration membranes is observed, despite the low molecular weight of less than 650 of the rhamnolipids (Fig. 11). The extent of surfactant retention depends on the pore diameter of the ultrafiltration, but not microfiltration membrane and is caused by membrane fouling and by the property of amphiphilic substances to associate and aggregate in high-molecular-weight micelles [68]. The aggregate size as well as the rhamnolipid retention depends on pH, temperature, ion milieu, shear stress and on the interaction between rhamnolipids and protein.

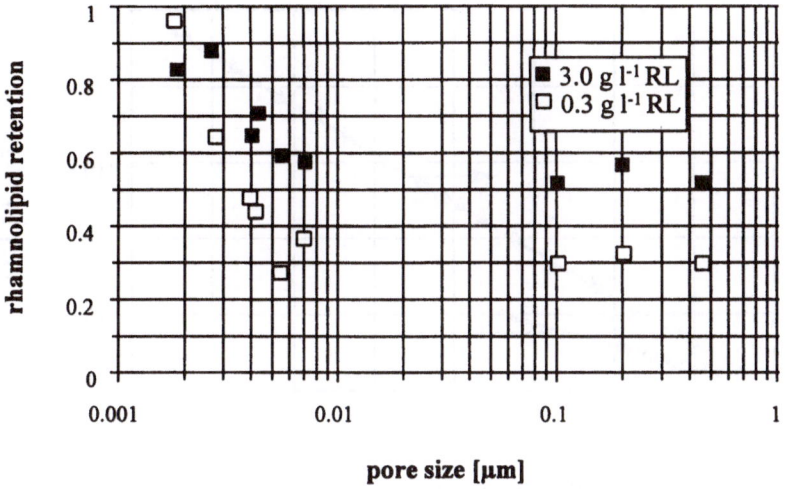

Fig. 11. Rhamnolipid retention as a function of membrane pore size and the rhamnolipid/protein concentrations. The ratio of the rhamnolipid and protein concentrations was 8 to 1. The protein was isolated from *P. aeruginosa* after cell disruption (taken from Gruber [68])

7.2 Gas Exchange

The gas exchange membranes are arranged as hollow fibers and consist of a polysulfone support layer which is coated with a thin polysiloxan layer. It is used to overcome the foam problem and to increase the process safety. An additional advantageous effect is the increased working volume of the bioreactor, because the large foam drainage zone in the headspace of the bioreactor is reduced, compared to systems with mechanical foam control [61, 62, 65]. Since carbondioxide is better soluble in the process broth than oxygen, the CO_2-flux through the membrane is much higher than the O_2-flux (Fig. 12). Therefore, the membrane area needed for oxygen entry is 50 times larger than the membrane area required for carbon dioxide removal [68]. Regarding the current membrane cost of 200 \$ m^{-2}, the membrane should be used exclusively for carbon dioxide removal, whereas oxygen can be added as pure gas directly to the system.

7.3 Downstream Processing

The ability to control the rhamnolipid retention by the filtration membrane provides the interesting perspective to integrate a part of the downstream processing into the production process itself. If the surface-active rhamnolipid is the target product, it may be advantagous to adjust the rhamnolipid retention in a way that a high amount of rhamnolipids is present in the cell-free permeate. These rhamnolipids can be isolated by precipitation and foam fractionating or

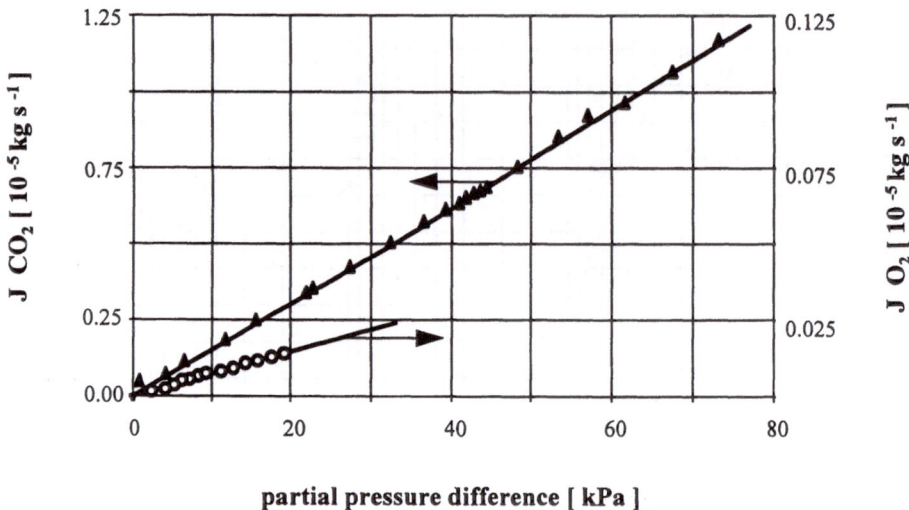

Fig. 12. Flux J of O_2 and CO_2 in correlation of the driving partial pressure. Membrane area = $0.182\,m^2$; Re(liquid) = 2030; Re(gas) = 512 (taken from Gruber [64])

by adsorption and subsequent chromatography on XAD-2 resins (66). If L-rhamnose is the target product, the rhamnolipid retention should be adjusted as high as possible, so that the bulk amount of rhamnolipid is available in the bleed and is reduced in volume. After hydrolysis of the rhamnolipids, L-rhamnose is accumulated in the aqeous phase of an emulsion and can be further purified by chromatography on ion-exchange resins and crystallization [70].

7.4 Operating Results

The process control and the composition of the corn-oil containing medium have been described in detail [65]. To investigate the influence of bleedrate variation on the productivity, the continuous process with cell recycling was carried out at a constant medium flow ($D = 0.1\,h^{-1}$) . By variation of the bleedrate, retention rates of 0.50, 0.65, 0.80, and 0.85 were employed. Retention rates above R = 0.85 were not investigated because the viscosity of the medium increased to inacceptable levels. Increasing retention rates resulted in an increase of rhamnolipid and biomass concentrations in the system, and the carbon source was utilized to a farther extent (Fig. 13). However, the volumetric productivity and product yield were similar at all retention rates (Table 9). Since the retention rate is an instrument to adjust the product stream between bleed and permeate, the product part increases from 3% at low retention rates to 32% at high retention rates.

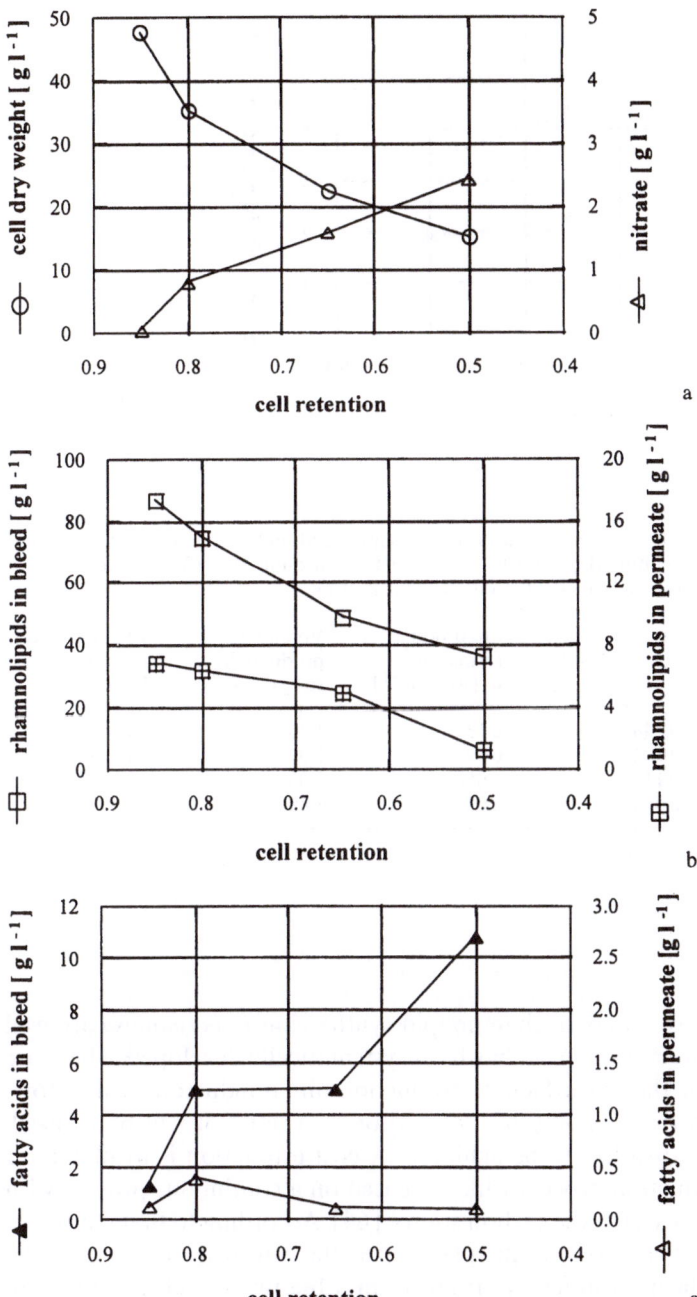

Fig. 13a–d. Influence of the retention on biomass and nitrate concentration **a**, on rhamnolipid concentration **b**, on substrate concentration **c**, and on protein concentration **d** in bleed and permeate during continuous cultivation of *P. aeruginosa*. The medium rate was kept constant at $D_0 = 0.1\,h^{-1}$, whereas the retention was varied. Medium M3 [65] containing $40\,g\,l^{-1}$ corn oil and growth-limiting concentrations of Fe was used; T = 33 °C, pH = 6.3

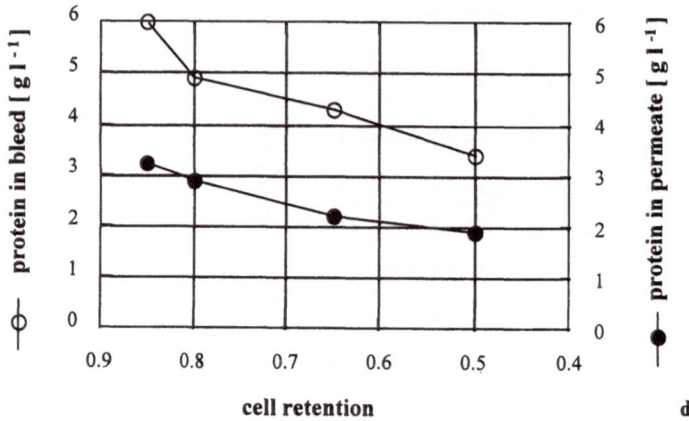

Table 9. Influence of the biomass retention on rhamnolipid yield and productivity. *P. aeruginosa* was grown in continuous culture ($D_0 = 0.1\ h^{-1}$; 33 °C; pH 6.3; medium M 3 [6] containing 40 g l^{-1} corn oil) with cell recycling and growth limitation by lack of Fe

Biomass retention R	Yield Y [g g^{-1}]	Sp. substrate consumption q_s [g g^{-1} h^{-1}]	Sp. product formation q_p [g g^{-1} h^{-1}]	Volumetric productivity rp [g l^{-1} h^{-1}]	Rhamnolipid in permeate [%]
0.50	0.413	0.248	0.131	1.86	3
065	0.419	0.174	0.090	2.01	16
0.80	0.484	0.118	0.056	2.00	26
0.85	0.473	0.083	0.039	1.88	32

8 Conclusions

The genetics and physiology of rhamnolipid synthesis in *P. aeruginosa* are well understood. A production process has been systematically developed. These are the prerequisites for the biosurfactant production on an industrial scale. However, it depends on the economy of such a process whether it will be realized, since a monetary return has to be achieved. A cost estimation is available for a rhamnolipid production process which is based on a continuous process with cell recycling and glucose as the carbon source [64]. According to that study, the rhamnolipid price depends on the plant size, the substrate costs and the retention rate. A cheap product will result from a big plant, a cheap substrate and a high retention. In the case of a free substrate, a medium flow of 100 m^3 d^{-1} and a retention above 0.8, the price of the raw rhamnolipid will be lower than 3 \$ kg^{-1}. The use of rhamnolipids as a source of rhamnose has great potential for an industrial application, since rhamnose represents an expensive compound for chemical and food industries. Using rhamnolipid as a competing surfactant on the world market would require a price below $2\ \$\,kg^{-1}$ [20]. Special permission would be required to make use of *P. aeruginosa* as a production strain because it is an opportunistic human-pathogen and is therefore not G.R.A.S. classified.

Acknowledgements. This project has been supported by grant No. 31–28763.90 from the Swiss National Science Foundation to PD Dr. J. Reiser. We wish to thank PD Dr. J. Reiser, Dr. A.K. Koch, PD Dr. O. Käppeli, Prof. Dr. B. Witholt, and Dr. I. Trösch IGB Stuttgart for their contributions.

9 References

1. Lang S, Wagner F (1987) Structure and properties of biosurfactants. In: Kosaric N, Cairns WL, Gray NCC (eds) Biosurfactants and Biotechnology, vol 25. M Dekker, New York, p 21
2. Syldatk C, Wagner F (1987) Production of biosurfactants. In: Kosaric N, Cairns WL, Gray NCC (eds) Biosurfactants and Biotechnology, vol 25. M Dekker, New York, p 89
3. Reiser J, Koch AK, Jenny K, Käppeli O (1989) Structure, properties, and production of biosurfactants. In: Obringer JW, Tillinghast HS (eds), Biotechnology for Aerospace Applications, Adv Appl Biotechnol vol 3. The Portfolio Publishing Co, The Woodlands, Tex, p 85
4. Georgiou G, Lin SC, Sharma, MM (1992) Bio/Technology 10: 60
5. Fiechter A (1992) Trends Biotechnol 10: 208
6. Desai JD, Desai AJ (1993) Production of biosurfactants In: Kosaric N (Ed) Biosurfactants: Production, Properties, Applications, vol 48. M Dekker, New York, Basel, Hong Kong, p 65
7. Jenny K (1990) Doctoral dissertation. Swiss Federal Institute of Technology, Zürich
8. Haferburg D, Hommel R, Claus R, Kleber HP (1986) Extracellular microbial lipids as biosurfactants. In: Fiechter A (Ed) Advances in Biochemical Engineering, vol 33. Springer, Berlin Heidelberg, New York p 53
9. Zhang Y, Miller RM (1992) Appl Env Microbiol 58: 3276
10. Lang S and Wagner F (1993) Biological activities of biosurfactants. In: Kosaric N (Ed) Biosurfactants: Production, Properties, Applications, vol 48. M Dekker, New York, p 251
11. Kosaric N, Gray, NCC, Cairns WL (1987) Biotechnology and the surfactant industry. In: Kosaric N, Cairns WL, Gray, NCC (eds), Biosurfactants and Biotechnology, vol 25. M Dekker, New York, p 1
12. Chakrabarty AM (1985) Trends Biotechnol 3: 32
13. Poremba K, Gunkel W (1991) Z Naturforsch 46c: 210
14. Müller-Hurtig R, Wagner F, Blaszczyk R, Kosaric N (1993) Biosurfactants for environmental control. In: Kosaric N (Ed), Biosurfactants: Production, Properties, Applications, vol 48. M Dekker, New York, p 447
15. Oberbremer A, Müller-Hurtig R (1989) Appl Microbiol Biotechnol 31: 582
16. Klekner V, Kosaric N (1993) Biosurfactants for cosmetics. In: Kosaric N (Ed) Biosurfactants: Production, Properties, Applications, vol 48. M Dekker, New York, p 373
17. Velikonja J, Kosaric N (1993) Biosurfactants in food applications. In: Kosaric N (Ed) Biosurfactants: Production, Properties, Applications, vol 48. M Dekker, New York, p 419
18. Greek BF (1991) Chemical and Engineering News 69: 25
19. Van Dyke MI, Lee H, Trevors JT (1991) Biotechnol Adv 9: 241
20. Mulligan CN, Gibbs BF (1993) Factors influencing the economics of biosurfactants. In: Kosaric N (Ed) Biosurfactants: Production, Properties, Applications, vol 48. M Dekker, New York, p 329
21. Jarvis FG, Johnson MJ (1949) J Am Chem Soc 71: 4124
22. Hisatsuka K, Nakahara T, Sano N, Yamada K (1971) Agr Biol Chem 35: 686
23. Itoh S, Honda H, Tomita F, Suzuki T (1971) J Antibiotics 24: 855
24. Syldatk C, Lang S, Wagner F (1985) Z Naturforsch Teil C, Biochem Biophys Biol Virol 40: 51
25. Hirayama T, Kato I (1982) FEBS Lett 139: 81

26. Rendell NB, Taylor GW, Somerville M, Todd H, Wilson R, Cole, J (1990) Biochem Biophys Acta 1045: 189
27. Hauser G, Karnovsky ML (1954) J Bacteriol 68: 645
28. Hauser G, Karnovsky ML (1957) J Biol Chem 224: 91
29. Hauser G, Karnovsky ML (1958) J Biol Chem 233: 287
30. Burger MM, Glaser L, Burton RM (1963) J Biol Chem 238: 2595
31. Burger MM, Glaser L, Burton RM (1966) Methods Enzymol 8: 441
32. Pazur JH, Shuey EW (1961) J Biol Chem 236: 1780
33. Kornfeld S, Glaser L (1961) J Biol Chem 236: 1791
34. Marumo K, Lindqvist L, Verma N, Weintraub A, Reeves PR, Lindberg AA (1992) Eur J Biochem 204, 539
35. Boulton CA, Ratledge C (1987) Biosynthesis of lipid precursors to surfactant production. In: Kosaric N, Cairns WL, Gray NCC (eds), Biosurfactants and Biotechnology, vol. 25. M Dekker, New York, p 47
36. Guerra-Santos LH, Käppeli O, Fiechter A (1986) Appl Microbiol Biotechnol 24: 443
37. Mulligan CN, Gibbs BF (1989) Appl Environ Microbiol 55: 3016
38. Mulligan CN, Mahmourides G, Gibbs BF (1989) J Biotechnol 12: 199
39. Guerra-Santos LH (1985) Doctoral dissertation, Swiss Federal Institute of Technology, Zürich
40. Ochsner U (1993) Doctoral dissertation, Swiss Federal Institute of Technology, Zürich
41. Holloway BW (1969) Bacteriol Rev 33: 419
42. Findlay JBC (1990) Purification of membrane proteins. In: Harris ELV, Angal S (eds) Protein purification applications, a practical approach, IRL Press, Oxford, p 59
43. Reiser J, Koch AK, Ochsner UA, Fiechter A (1993) Genetics of surface-active compounds. In: Kosaric N (Ed) Biosurfactants: Production, Properties, Applications, vol. 48. M Dekker, New York, p 231
44. Cosmina P, Rodriguez F, de Ferra F, Grandi G, Perego M, Venema G, van Sinderen D (1993) Mol Microbiol 8: 821
45. Nakano MM, Zuber P (1993) J Bacteriol 175: 3188
46. Nakano MM, Corbell N, Besson J, Zuber P (1992) Mol Gen Genet 232: 313
47. Koch AK, Käppeli O, Fiechter A, Reiser, J (1991) J Bacteriol 173: 4212
48. Siegmund I, Wagner F (1991) Biotechnology Techniques 5: 265
49. Koch AK (1992) Doctoral dissertation, Swiss Federal Institute of Technology, Zürich
50. West SEH, Iglewski BH (1988) Nucl Acids Res 16: 9323
51. Ochsner UA, Fiechter A, Reiser J (1994) J Biol Chem 269: 19787
52. Ochsner UA, Koch AK, Fiechter A, Reiser J (1994) J Bacteriol 176: 2044
53. Ochsner UA, Reiser J (1995) Proc Natl Acad Sci USA, in press
54. Henikoff S, Wallace JC, Brown JP (1989) Methods Enzymol 183: 111
55. Swift S, Bainton, NJ, Winson MK (1994) Trends Microbiol 2: 193
56. Kustu S, Santero E, Keener J, Popham D, Weiss D (1989) Microbiol Rev 53: 367
57. Ochsner UA, Reiser J, Fiechter A, Witholt B (1995) Appl. Environ. Microbiol. (submitted)
58. Davison J, Heusterspreute M, Chevalier, N, Ha-Thi V, Brunel F (1987) Gene 51: 275
59. De Lorenzo V, Eltis L, Kessler B, Timmis KN (1993) Gene 123: 17
60. Huijberts GNM, Eggink G, De Waard P, Huisman GW, Witholt B (1992) Appl Env Microbiol 58: 536
61. Syldatk C (1984) Doctoral dissertation, TU Braunschweig
62. Linhardt RJ, Bakhit R, Daniels L, Mayerl F, Pickenhagen W (1989) Biotechnol Bioeng 33: 365
63. Reiling HE, Thanei-Wyss U, Guerra-Santos LH, Hirt R, Käppeli O, Fiechter A (1986) Appl Environ Microbiol 51: 985
64. Gruber T (1991) Doctoral dissertation, U Stuttgart
65. Hembach T (1995) Doctoral dissertation, U Hohenheim
66. Matulovic U (1987) Doctoral dissertation, TU Braunschweig
67. Robert M, Mercadé ME, Bosch ME, Parra JL, Espuny MJ, Manresa MA, Guinea J (1989) Biotechnol Lett 11: 871
68. Gruber T, Chmiel H, Käppeli O, Sticher P, Fiechter A (1993) Integrated process for continuous rhamnolipid production. In: Kosaric N (Ed) Biosurfactants: Production, Properties, Applications, vol 48. M Dekker, New York, p 157
69. Johnson MK, Boese-Marrazzo D (1980) Infect Immun 29: 1028
70. Wullbrandt D, Giani C, Mixich J, Kunz M, Rapp KM, Vogel M (1994) 12. DJB (in press)

Microbial Carotenoids

Eric A. Johnson and William A. Schroeder
University of Wisconsin, Departments of Food Microbiology and Toxicology,
1925 Willow Drive, and Bacteriology, 1550 Linden Drive, Madison,
Wisconsin 53706, USA

Carotenoids occur universally in photosynthetic organisms but sporadically in nonphotosynthetic bacteria and eukaryotes. The primordial carotenogenic organisms were cyanobacteria and eubacteria that carried out anoxygenic photosynthesis. The phylogeny of carotenogenic organisms is evaluated to describe groups of organism which could serve as sources of carotenoids. Terrestrial plants, green algae, and red algae acquired stable endosymbionts (probably cyanobacteria) and have a predictable complement of carotenoids compared to prokaryotes, other algae, and higher fungi which have a more diverse array of pigments. Although carotenoids are not synthesized by animals, they are becoming known for their important role in protecting against damage by singlet oxygen and preventing chronic diseases in humans. The growth of aquaculture during the past decade as well as the biological roles of carotenoids in human disease will increase the demand for carotenoids. Microbial synthesis offers a promising method for production of carotenoids.

Advances in Biochemical Engineering
Biotechnology, Vol. 53
Managing Editor: A. Fiechter
© Springer-Verlag Berlin Heidelberg 1995

List of Symbols and Abbreviations

ψ (psi)	Acyclic
β (beta)	Cyclohexene
ε (epsilon)	Cyclohexene
λ (lambda)	Lambda (wavelength)
CHEF	Contour-clamped Homogenous Electric Field electrophoresis
C/N	Carbon to nitrogen ratio
DCW	Dry cell weight
EMS	Ethyl methane sulfonate
ER	Endoplasmic reticulum
E/Z	*cis/trans* Stereoisomers
FACS	Fluorescence activated cell sorting
HDCO	3-Hydroxy-3',4'-didehydro-β, ψ-caroten-4-one
HMG-CoA-reductase	3-Hydroxy-3-methylglutaryl-coenzyme A reductase
IUPAC	International Union of Pure and Applied Chemistry
NAD(P)	β-Nicotinamide Adenine Dinucleotide (Phosphate)
NTG	*N*-Methyl-*N*'-nitro-*N*-nitrosoguanidine
PG	*n*-Propyl gallate
RFLP	Restriction Fragment Length Polymorphism
SHAM	Salicylhydroxamic acid
$^1O_2$	Singlet oxygen
μ	Specific growth rate
Yx/s	Yield of cells on substrate

1 Introduction

Although carotenoids occur widely in nature, they are by no means unified in biological organisms. Carotenoids show much diversity in natural distribution, structure, and function. The primary functions of carotenoids in photosynthetic organisms are as light harvesting accessory pigments and as protectants against porphyrin-catalyzed photooxidations [1–3]. Carotenoids have the specific and catalytic ability to quench singlet oxygen ($^1O_2$) generated from metabolism or by the reaction of triplet-state porphyrins with oxygen. Even in nonphotosynthetic organisms, a universal role of carotenoids is to sequester toxic oxygen species from metabolism or the environment, most specifically $^1O_2$. The diversity of carotenoids has evolved in relation to functional requirements, but they have apparently been deposited in animals by availability in the food chain. Like hormones and sterols, they initially occurred by necessity in some organisms, but apparently in others by gratuity.

Carotenoids were initially synthesized by anoxygenic phototrophic bacteria and oxygenic photosynthetic prokaryotes (cyanobacteria) [4–6]. In eukaryotes, the capacity to synthesize carotenoids was acquired by symbiosis with bacteria or with eukaryotic chloroplasts [7–9]. In terrestrial plants, and red and green algae, stable symbioses were established with cyanobacteria or related prochlorophytes. The origin of carotenoids in non-red and non-green algae as well as the fungi appears polyphyletic and possibly evolved independently and repeatedly in different taxonomic groups [7, 8, 10]. Carotenoids are not synthesized by animals although carotenes can be altered in some animals by oxidation or isomerization and by conversion to vitamin A and retinoids.

Since many animals of agricultural importance contain carotenoids and require them as nutrients, these pigments are industrially significant as components in animal feeds. Carotenoids are required as feed supplements in the poultry industry and in aquaculture of fishes and crustaceans. Besides providing nutrition and possibly disease resistance, carotenoids give brilliant pigmentation and aesthetic value to crustacea, animals, birds, and their ova. Recent studies have suggested that carotenoids have health benefits in humans and animals by preventing or delaying some chronic diseases including cancer, arteriosclerosis, cataracts, and other maladies. These developments have contributed greatly to the industrial and medical interest in carotenoids. Traditionally, carotenoids for agriculture and food uses were obtained by extraction of natural plant sources (e.g. marigold flowers or corn) and by chemical synthesis. During the past ten years, much interest and effort has been devoted to designating and developing microbial sources for industrially important carotenoids. Microbial synthesis could provide a significant proportion of the pigments used in terrestrial agriculture and marine aquaculture, and potentially as human nutrients or "nutraceuticals".

2 Nomenclature and Structure

Carotenoids are isoprenoids containing a characteristic polyene chain of con-jugated double bonds. The two general groups of pigments are the hydrocar-bons (carotenes) and oxygenated derivatives (xanthophylls). Carotenoids are formed from two geranylgeranyl precursors by head to head condensation which gives a symmetrical acyclic $C_{40}H_{56}$ basic structure (Fig. 1) [11]. The basic molecule is arranged such that the two central methyl groups are in a 1,6 position. About 600 natural carotenoids are derived from the basic molecule by various enzymatic chemical processes. About 370 of the naturally occurring carotenoids have chiral centers, and in some cases both the R and S configura-tions have been isolated. In addition, *cis/trans* (*E/Z*) -isomerism occurs widely in carotenoids. Normally carotenoids exist in nature as the Z form, but several exceptions exist [11].

Rules for nomenclature of carotenoids have been recommended by IUPAC [12, 13]. Trivial names will be used in this review; the semisystematic names have been tabulated [14]. It should be kept in mind that several of the carotenoids cited in this review were identified by absorbance and chromato-graphic characteristics alone. The structure of these pigments should be con-firmed (including the stereochemistry) by high resolution MS, NMR and CD spectroscopy.

3 Optical Properties of Carotenoids

The region of the carotenoid molecule related to light absorption is the polyene chain consisting of a series of conjugated double bonds. The wavelength of maximum absorption is directly related to the number of conjugated double

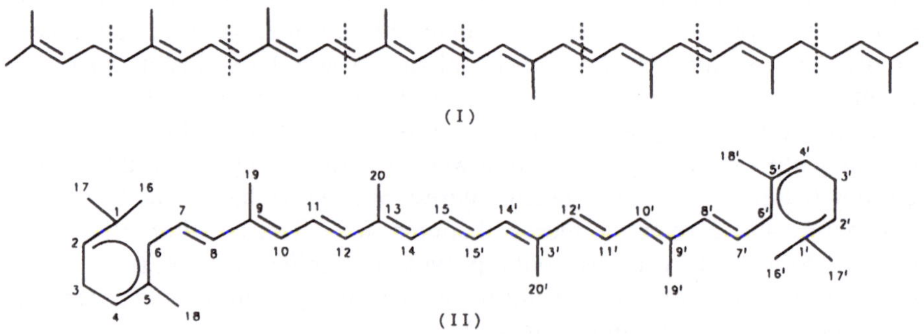

Fig. 1. I: Basic acyclic $C_{40}H_{56}$ carotene structure from which all carotenoids are derived; **II:** Structure and numbering of carotene stem molecule

bonds. The absorption maximum can be estimated from the equation λ_{max}(nm) = 300.5 + 65.5 × n where n is the number of conjugated double bonds present. Each additional double bond increases the λ_{max} by 7–35 nm (5–9 nm if the double bond is located in a ring). This is referred to as a bathochromic shift. The addition of carbonyls conjugated to the polyene chain has two effects on optical properties. The first is to increase the λ_{max} by ~28 nm for a first carbonyl in the polyene chain. Additional carbonyls either in the polyene chain or in a ring structure will increase the λ_{max} by 1–9 nm. Normally, the polyene chain has a characteristic absorbance spectrum with three peaks. This characteristic shape is referred to as having high persistence. The introduction of the carbonyl results in loss of persistence and loss in fine structure. This yields a rounded, symmetrical peak of absorption. Other chemical modifications of the carotenoid structure, with the exception of epoxides, have little effect on absorbance characteristics, since they do not sterically interfere with the polyene chain. Epoxides decrease the λ_{max} by 10–20 nm (a hypsochromic shift). The 9 terminal carbons of each end of microbial xanthophylls are usually arranged in one of 3 configurations. In the acyclic ψ configuration, steric hindrance is minimized, yielding a longer wavelength chromophore. When the end group is cyclized to a β group, steric hindrance is increased, shortening the λ_{max}. In the third common end group, the ε configuration, the terminal double bond is displaced one position relative to the β end group double bond. This bond is no longer conjugated to the chromophore and therefore gives a hypsochromic shift.

4 Biological Properties of Carotenoids

4.1 Evolutionary Origins of Carotenoids

Advances in microbial phylogeny, particularly by 16s rRNA sequencing [15, 16] and by analysis of conserved gene and protein sequences [7, 17, 18] have improved our understanding of the evolutionary origins of prokaryotes and eukaryotes. Molecular comparisons have demonstrated three primary taxa of life on Earth, the Eubacteria, Archaebacteria, and Eukaryota (Protista) [9, 15] [Fig. 2]. Of the three groupings, carotenoids are known to be produced in eubacteria including all phototrophic bacteria and sporadically in heterotrophic bacteria, but by only one group of archaebacteria, the halobacteria. Carotenoids are produced universally in photosynthetic algae and plants, and by some fungi. They are not found in strictly anaerobic eubacteria and archaebacteria nor are they synthesized by animals.

Xanthophylls, like sterols [19], probably evolved in relation to the appearance of oxygen. The evolution of life consisted of two main periods: (1) the age of bacteria 3500–1700 million years ago and (II) the age of eukaryotes from 1700 million years ago to the present [20]. Carotenes and xanthophylls found in

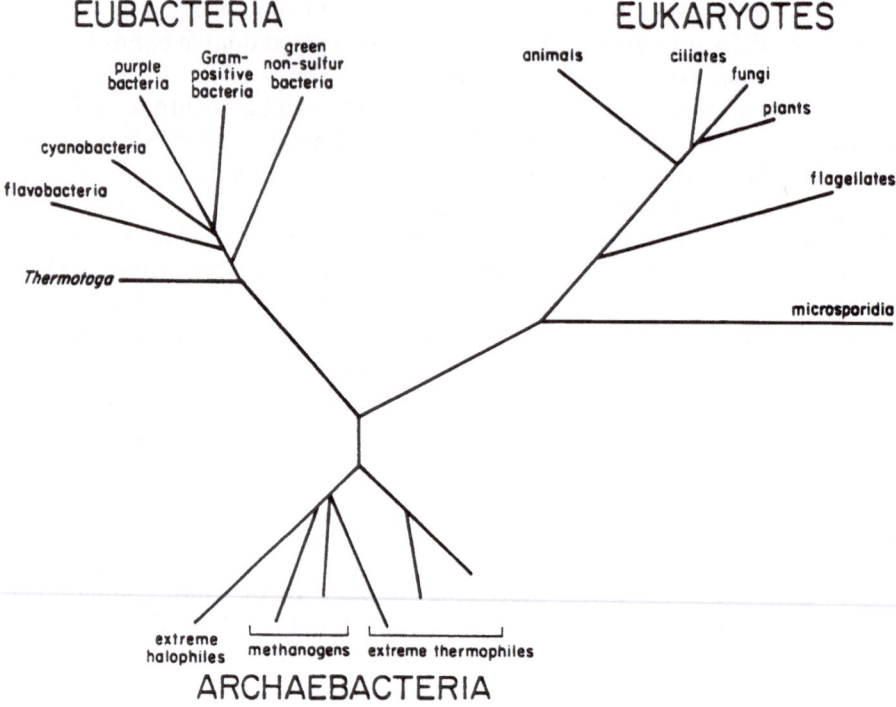

Fig. 2. Universal phylogenetic tree determined from rRNA sequences [from Ref. 16]

anoxygenic photosynthetic bacteria probably evolved in the bacterial epoch and were adapted as accessory pigments in anoxygenic photosynthesis. From 16s rRNA-based phylogeny, 12 major groups of evolutionary lineages of bacteria diverged from each other at essentially the same time, which corresponded to the onset of photosynthesis [5, 21]. Carotenoid biosynthesis probably occurred initially in anoxygenic photosynthetic bacteria. These organisms, which include the photosynthetic purple bacteria, green bacteria, and heliobacteria, contain bacteriochlorophylls and group-specific aliphatic and monocyclic carotenoids as the primary light harvesting pigments [22–24]. Ancient rock sediments formed about 3.5 billion years ago in Australia and South Africa showed evidence of at least four types of filamentous prokaryotes living in stromatolites in shallow lakes [25, 26]. The organisms resemble filamentous mat-forming contemporary cyanobacteria and anoxygenic photosynthetic bacteria [5, 25]. The ancestral carotenogenic cell may have resembled the present day *Heliobacterium chlorum* (gram-positive line) [27], which contains the simple alicyclic carotene neurosporene as its primary carotenoid. This organism is strictly anaerobic and has the simplest known porphyrin-based photosynthetic reaction center [28]. The high O_2-sensitivity of this bacterium is related to its low content of neurosporene [27] which is a relatively poor quencher of $^1O_2$. As the

Table 1. Carotenoid groups and characteristic features from anoxygenic phototropic bacteria Adapted from [23, 30]

Group	Cartenoid groups	Major pigments	Structural characteristics
1	Normal spirilloxanthin	Neurosperene, lycopene, rhodopin, spirilloxanthin	Aliphatic, methoxyl groups
2	Alternative spirilloxanthin	Chloroxanthin, spheroidene, speheroidene, spirilloxanthin	Aliphatic, methjoxyl groups keto group
3	Okenone series	Okenone	Aromatic, keto and methoxyl groups
4	Rhodopinal series (related to group 1)	Lycopene, lycopenal, lycopenol, rhodopin, rhodopinal, spirilloxanthin	Aliphatic, Aldehyde (C_{20}), methoxyl and hydroxyl groups
5	Chlorobactene	Chlorobactene, isorenieratene, β-carotene, γ-carotene	Aromatic and cyclohexane rings
6	Chloroflexus	Lycopene, γ-carotene, echinenone, β-carotene, carotenoid glycosides	Cyclohexene ring, keto groups, glycosyl groups.

anoxygenic phototrophic bacteria evolved, there was an accompanying diversification in the complement of carotenoids (Table 1). Cell suspensions of these organisms are striking, showing purple-violet, purple, red, orange-brown, yellowish-brown, and orange-green colors [29]. The colors are due to carotenoids produced by the various groups reflecting quite different pathways of biosynthesis in the later stages (Table 1) [30]. Many of the pigments contain oxygen functions (especially ketone groups at the C_4 and C_{20} positions), which is interesting since it is believed that xanthophylls commonly derive from reaction of molecular oxygen by mixed-function oxidase (e.g. flavoenzymes or cytochrome P-450) enzyme systems, and oxygen groups would not be expected to be involved in biosynthesis in strictly anaerobic environments.

The differences in carotenoid structures in the anoxygenic photosynthetic bacteria is probably necessary for the optimum exploitation of the spectra of light that filters through microbial assemblages or mats [5, 24]. The carotenoids in the bacteria are located in chromatophores or "grana" in the extensive lamellar, tubular or vesicular membrane systems continuous with the cytoplasmic membrane. The purple nonsulfur bacteria are the most diverse group metabolically, in internal membrane structure, and in carotenoid composition. Most representatives grow under microaerophilic to aerobic conditions in the dark as chemoorganotrophs, and under these conditions the synthesis of photosynthetic pigments including carotenoids is repressed. The green sulfur bacteria, which contain chlorobactene or isorenieratene, are strictly anaerobic and obligately phototrophic [31]. Of the anoxygenic phototrophs, *Chloroflexus* is unusual in its pigment content producing β-carotene and hydroxy-γ-β-carotene glucoside anaerobically and echinenone and 4-keto-γ-carotene glucoside (myxobactone) in aerobic conditions [32]. *Thermomicrobium roseum* is a green nonsulfur bacterium originally isolated from the effluent of a hot spring in Yellowstone National Park [33] which also has an unusual complement of carotenoids. The primary carotenoids in *T. roseum* appear to be torulene and 3,4-dehydrolycopene [33]. The carotenoids found in the anoxygenic groups are listed in Table 1.

An extremely important event in the evolution of carotenoids was the development of photosystem II, which allowed the utilization of H_2O as an electron donor in photosynthesis and release oxygen as a product. The geological record particularly the changes in iron-banded layers indicates that substantial quantities of oxygen were present in the atmosphere about 1.7 billion years ago [34]. Oxygenic photosynthesis occurred in bacteria similar to present day cyanobacteria and prochlorophytes [6, 18]. In cyanobacteria, the major light-harvesting pigments are chlorophyll a and phycobiloproteins. The carotenoids in the cyanobacteria are structurally different than the pigments found in the anoxygenic phototrophic bacteria. Cyanobacteria form cyclic ketocarotenoids including echinenone, canthaxanthin, and in some species caloxanthin and nostoxanthin. They also synthesize glycosylated xanthophylls including myxoxanthophyll and oscillaxanthin. The carotenoids apparently are not integrally involved in light harvesting but may provide protection against photooxidation. The second order of oxygenic photosynthetic bacteria are the Prochlorales. This group has an unusual pigment content; it contains chlorophylls a and b but lacks phycobiloproteins as light harvesting pigments [35, 36]. The prochlorophytes contain chlorophylls a and b, β-carotene, zeaxanthin, cryptoxanthin, echinenone, and mutatochrome as the primary light harvesting pigments [22, 37, 38]. The pigment composition of the prochlorophytes appears to more closely resemble the pigment composition of green algae and higher plants except for a lack of lutein and has been considered as the evolutionary ancestor of chloroplasts [8, 35]. A novel prochlorophyte was reported to contain chlorophyll a and b, and the carotenoids α-carotene and zeaxanthin (typical plant pigments) but they lack β-carotene, lutein and prasinoxanthin [39]. The presence of α-carotene was puzzling because procaryotes are not known to synthesize ε rings, but its presence suggests a relation to plant and green algal chloroplasts. The prochlorphytes have been proposed as being a descendant of ancestral chlorophytes, the free-living ancestor of chloroplasts, or a deviant cyanophyte derivative [40].

The evolution of oxygenic photosynthesis about 3 to 2.5 billion years ago [6] had enormous consequences for biochemical evolution: O_2 began to accumulate and the atmosphere changed from reducing to oxidizing. As O_2 gradually accumulated, the fossil record suggests that there was an enormous burst in the rate of evolution leading to the formation of eukaryotes with organelles and eventually to multicellular plants and animals [18]. Terrestrial plants and probably the red algae (Rhodophyta) evolved their photosynthetic capability through endosymbiosis with bacteria resembling present day cyanobacteria [9, 41–45]. The common ancestor of cyanobacteria and purple bacteria was probably an O_2-evolving photoautoroph, related to the purple bacteria, which had earlier lost its capacity to use water as an electron donor [6, 46].

In certain lineages of eukaryotes, including some orders of algae and fungi, multiple endosymbiotic events with cyanobacteria or with eukaryotic chloroplasts expanded organelle diversity [47]. The nonphotosynthetic carotenogenic bacteria probably evolved from photosynthetic bacteria by loss of photosynthesis [6, 16, 21].

The gradual appearance of oxygen probably had tremendous consequences for the diversification of carotenoid structure. Like sterols, which primarily evolved in parallel with the appearance of oxygen [6], the diversity of carotenoids undoubtedly expanded and adapted to special functions. The reaction of carotenoids with oxygen-dependent enzyme systems has created diversity in structure. During the transition from anaerobic to an O_2-containing environment on earth, the carotenoids could have played a role as a rudimentary antioxidant system for coping with reactive oxygen species.

4.2 Carotenoid Structures and Phylogeny

About 600 carotenoids structures are now known to occur naturally [11]. There is considerable similarity in the carotenes of various bacteria, plants, and protists suggesting a universal early pathway (at least up to phytoene or neurosporene), but divergence occurred in later steps of carotene synthesis and many variations exist in the structure of xanthophylls (Table 2). The diversity of carotenoid structures is greatest in the algae and bacteria, less diverse in fungi, and least variable in terrestrial plants. The diversity of photopigment systems is of ecological significance for the harvesting of light energy in mat ecosystems in the algae and prokaryotes [48, 49] and probably for inactivating toxic oxygen species in heterotrophic species.

4.2.1 Bacteria

The distribution of carotenoids has evolved in several groups of prokaryotes. This distribution is reviewed in light of recent advances in our understanding of a natural phylogeny [16, 18] and the need to identify potential new sources of industrially important carotenoids.

Within the two kingdoms of bacteria, only the halobacteria in the archaebacteria are known to contain carotenoids, whereas many species of eubacteria produce carotenoids. The bright red color of cell suspensions of halobacteria e.g. (*Halobacterium*, and other halotolerant species of archaebacteria) is due to carotenoids. C_{40} carotenoids are present including β-carotene, but the C_{50} bacterioruberins are the primary pigments. These are alicyclic pigments with 2 to 4 hydroxyl groups. Carotenes in *Halobacterium* include the usual series of intermediates, except that the pathway differs from the one found in plants by the formation of *trans*-isomers instead of the *cis*-isomers [50]. Carotenoids

Table 2. Representative carotenoids reported in bacteria, plants, algae and fungi, Adapted and expanded from [73, 87, 88])

Phylogenetic group	Representative Carotenoids
A. Eubacteria (classified according to Ref. 5)	
Purple phototrophic bacteria (see Table 1)	Anaerobic: Neurosporene, rhodopin, lycopene, spirilloxanthin Aerobic: spheroidene, hydroxy-spheroidene, okenone
Green sulfur bacteria	
Chlorobium	Chlkorobactene (green strains) β-isorenieratene, isorenieratene
Green nonsulfur bacteria	
Chloroflexus	β-Carotene, γ-carotene
Herpetosiphon	Myxobacton fatty esters
Cyanobacteria	
Synechococcus	β-Carotene, echinenone, 3-hydroxyechinenone, zeaxanthin, cryptoxanthin, caloxanthin, nostoxanthin. Glycosides myxoxanthophyll, and oscillaxanthin also occur.
Procholorophytes	β-Carotene, zeaxanthin
Purple nonphotorophic bacteria	
Myxococcus	Myxobacton and glycosides, 4-keto torulene
Stigmatella	Myxobacton and acyl esters
Sorangium	γ-Carotene, carotenoid glucoside acyl esters
Chondromyces	γ-Carotene, carotenoid glucoside acyl esters
Bradyrhizobium	2,3,2'3'-Ditrans-tetrahydroxy-β-carotene and 2,3,2'3'-ditrans-tetrahydorxy-β-caroten-4-one
Pseudomonas	Rhodoxanthin, nostoxanthin, decaprenoxanthin
Xanthomonas	Brominated arylpolyenoic esters
Spirchetes	
Spirochaeta	1',2'-Dihydro-1'-hydroxytorulene 4-keto-1',2'-dihydro-1'-hydroxytorulene
Bacteroides-Flavobacterium-Cytophagas	
Flavobacterium	Zeaxanthin, astaxanthin (?). Also C_{45} and C_{50} carotenoids.
Flexibacter	Flexirubin, Chloroflexirubin, flexixanthin, deoxyflexixanthin
Cytophaga	Zeaxanthin, flexirubins, chloroflexirubins, flexixanthin, deoxyflexixanthin
Saprospira	Saproxanthin
Deinococcus-Thermus	
Deinococcus	β-Carotene, keto-carotenoids
Thermus	α-Carotene (?), γ-carotene, unidentified polar carotenoids
Gram-positive bacteria-high GC	
Streptomyces	Leporotene and hydroxy derivatives
Nocardia	Lycopene, γ-carotene, 4-keto-γ-carotene, carotenoid glycosides.
Corynebacterium	C_{45} and C_{50} carotenoids, glycoside of decapreno-xanthin
Brevibacterium	Canthaxnthin, β-carotene, astaxanthin
Mycobacterium	Isorenieratene (aromatic carotene), β-carotene, carotenoid glycosides
Micrococcus	Sarcinaxanthin, canthaxanthin, echinenone
Gram-positive bacteria-high GC	
Bacillus (alkalophilic)	β-Carotene, C_{50} carotenoids
Staphylococcus	C_{30} apocarotenoids
Sarcina	Lycopene, β-carotene, zeaxanthin, C_{50} carotenoids
Enterococcus	C_{30} carotenoids

Table 2. (*continued*)

Phylogenetic group	Representative Carotenoids
B. Archaebacteria	
Extreme halophiles-*Halobacterium*	β-Carotene, C_{50} xanthophylls-derivatives of acyclic bacterioruberin. Possess retinal protein complex, bacterorhodopsin
Hyperthermophiles-Sulfolobus	
C. Protista	
Chlorophyta (green alage)	β-Carotene, lutein, violaxanthin, neoxanthin, zeaxanthin, loroxanthin, siphonoxanthin, astaxanthin
Euglenophyta (euglenids)	Diadnoxanthin, diatoxanthin, heteroxanthin, eutreptiellanone (oxabicycloheptane carotenoid), siphonein, neoxanthin.
Phaeophyta (brown algae)	Fucoxanthin, violoaxanthin, acyl-fucoxanthins
Pyrrophyta (dinoflagellates)	Peridinin (C_{37}), diadinoxanthin, acyloxyfucoxanthins, gyroxanthins
Chrysophyta (golden-brown algae, diatoms)	Fucoxanthin, other acetylenic carotenoids
Rhodophyta (red algae)	α-Carotene, β-carotene, lutein, zeaxanthin
Cryptophyta	Alloxanthin, crocoxanthin, monadoxanthin
D. Higher plants	
Chloroplasts	α-Carotene, β-carotene, lutein, β-cryptoxanthin, Violaxanthin, neoxanthin
Chromoplasts (species specific)	Lycopene, β-carotene, β-cryptoxanthin, zeaxanthin, apocarotenoids, 5,6-epoxy- and 5,8-epoxycarotenoids, capsanthin
E. Fungi	
Myxomycota	
Myxomycetes (slime molds)	γ-Carotene, β-carotene, 3,4-didehydrolycopene, neurosporaxanthin
Eumycota	
Mastigomycotina	
Chytridiomycetes	γ-Carotene, lycopene, β-carotene
Oomycetes	
Zygomycotina	
Zygomycetes	β-Carotene
Ascomycotina	
Hemiascomycetes	None known
Plectomycetes	γ-Carotene, β-carotene, lycopene
Pyrenomycetes	γ-Carotene, β-carotene, lycopene, torulene, lycoxanthin, neurosporaxanthin
Discomycetes	β-Carotene, γ-carotene, 3,4-didehydrolycopene, torulene, phillipsiaxanthin, aleuriaxnthin, plectaniaxanthin
Loculoascomycetes	None known?
Basidiomycotina	
Hymenomycetes	γ-Carotene, β-carotene, cryptoxanthin, canthaxanthin, ataxanthin (?)
Gasteromycetes	γ-Carotene, β-carotene
Teliomycetes	β-Zeacarotene, γ-carotene, β-carotene, torulene, cryoptoxanthin
Basidiomycetous yeasts	β-Carotene, torulene, torularhodin
	β-carotene, HDCO, echinone, astaxanthin
Deuteromycotina	γ-Carotene, β-carotene, torulene

Many of the designated carotenoids were identified based on chromatographic and visual absorbance; it would be valuable to confirm the structures by modern spectroscopic methods

protect the halophilic cells against damage by bright light [51]. The halobacteria also have evolved energy-producing bacteriorhodpsin [52], which contains
retinal derived from carotenoids as a prosthetic group. Bacteriorhodopsin can
occupy as much as 50% of the protein of the cell membrane.

The eubacteria possess a rich variety of carotenoids and the pigments occur in
specific phylogenetic groups. From the sequences of nucleotides of fragments of
16s rRNA, 12 distinct eubacterial groups have been delineated [16, 53] [Fig. 3].
These include: (1) cyanobacteria, (2) green sulfur bacteria, (3) multicellular
filamentous green nonsulfur bacteria (*Chloroflexus*) and chemotrophic related
bacteria, (4) phototrophic purple bacteria, (5) planctomyces, (6) spirochetes, (7)
flavobacteria-*Bacteroides*, (8) chlamydiae, (9) deinococci-*Thermus*, (10) Gram-
positive bacteria, (11) *Thermotoga*, and (12) *Aquifex* [16]. Of the 12 groups,
carotenoids occur in the photosynthetic bacteria (groups 1–4), spirochetes,
flavobacteria-cytophages, deinococci, and gram-positive lineages. The carotenoids of the phototrophic bacteria (in groups 1–4) were described in the
previous section. Carotenoids occur sporadically in the nonphotosynthetic
bacteria. They are rarely found in strict anaerobes except for the phototrophic
bacteria.

Carotenogenic purple bacteria include pseudomonads and myxobacteria
[54, 55]. Although carotenoids are not common in rhizobia, the species *Brady-
rhizobium* contains hydroxy-derivatives of nostoxanthin [56]. The same pig-

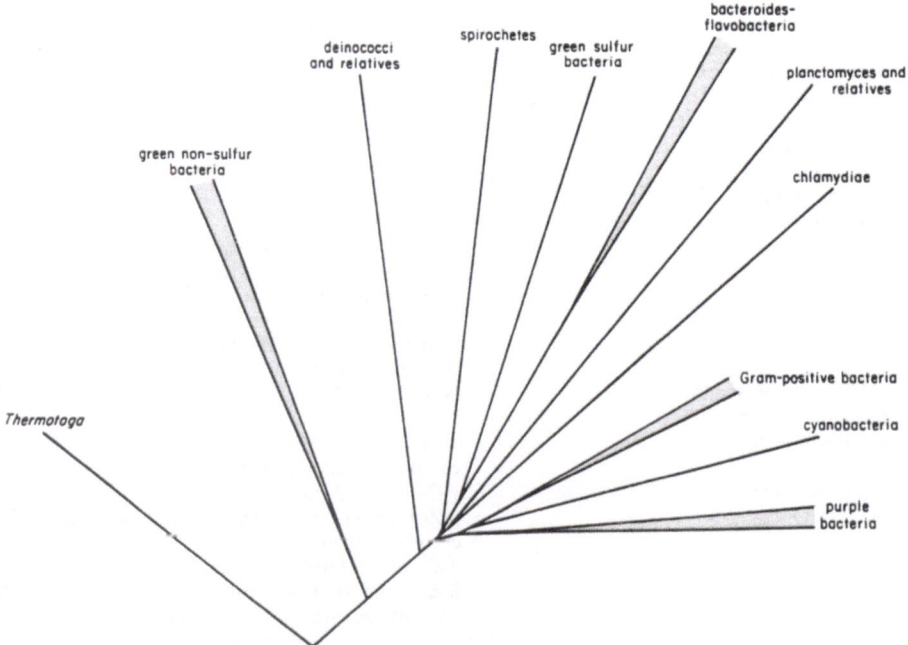

Fig. 3. Eubacterial phylogenetic tree from comparison of 16s rRNA sequences [from Ref. 16]

ment also occurs in *Pseudomonas rhodos* [57]. Carotenoids of other pseudomonads are rather poorly characterized. A pigment similar to 3,5,3′-trihydroxy-5,6-dihydro-β-carotene was reported in *P. echinoides*, and mutants that accumulated lycopene, γ-carotene, or phytoene were obtained [54]. Nostoxanthin was identified as the primary carotenoid in several *Pseudomonas* species [58]. The obligate methylotrophic bacteria, which utilize methane and methanol as sole carbon source, are pink colored apparently due to the presence of a ketocarotenoid associated with the outer membrane [59]. The aquatic organism *Azospirillum* produces unidentified carotenoids under aerobic conditions [60]. The parasite *Bdellovibrio* also produces an unidentifed carotenoid [61]. Carotenoids are relatively uncommon in the family Enterobacteriaceae which includes many human pathogens. However, *Erwinia* is a yellow colored plant-associated enteric bacterial genus whose species contain β-carotene, β-cryptoxanthin, and zeaxanthin and glycosides as its major carotenoids [62]. *Xanthomonas* species produce unique carotenoids containing an aryl polyene acid chain with bromine covalently attached [63]. One interesting group, isolated recently and included in the genera *Roseobacter* and *Erythrobacter*, possesses unidentified carotenoids whose synthesis is stimulated by light [64].

Carotenoids are common in spirochetes which produce the torulene derivatives 4-keto-1′, 2′-dihydro-1′-hydroxytorulene and 1′, 2′-dihydro-1′-hydroxytorulene. The flavobacteria lineage is an important source of carotenoids. β-Carotene and zeaxanthin have been isolated from flavobacteria. In addition, the C_{45} and C_{50} pigments nonaprenoxanthin and 11′, 12′-didehydrononaprenoxanthin, an apocarotenoid, have been isolated from *Flavobacterium dehydrogenans* [54]. Phytoene in *F. dehydrogenans* is all *trans*, in contrast to the *cis*-isomer found in most bacteria [65]. Other *Flavobacterium* spp. produce carotenoids and a recent isolate from Yellowstone Park and a marine isolate from Japan apparently produce high levels of astaxanthin which is glycosylated and possibly secreted [66].

The gliding bacteria (Cytophagales) produce several carotenoids which occur in the cytoplasm, while the novel flexirubin pigments are bound to the membrane [67]. The carotenoids include zeaxanthin and the monocyclic ketocarotenoids, flexixanthin and deoxyflexixanthin [67]. The flexirubins, which have been valuable in chemosystematics, have a chromophore of omega-phenyloctaenic acid which is connected via an ester linkage to a resorcinol derivative. The basic chemical structure can be modified by the introduction of methyl and chlorine on the omega-phenyl ring. The carotenoids in *Myxococcus fulvus* increased in the presence of light and in late logarithmic phase of growth to reach levels of 0.06% of the dry weight [68]. The carotenoids are mainly located in the cytoplasmic membrane, comprising 0.14% of its weight. With the exception of *Nannocystis exedens*, the main pigments are carotenoid glycosides with a fatty acid attached to the sugar. Myxobacton, the primary pigment in many myxobacteria, is characteristic for this taxonomic subgroup [69]. Other carotenids include γ-carotene, lycopene, and 4-ketotorulene. The only known role of carotenoids in myxobacteria is to provide protection against photooxidation [69]. The aquatic gliding bacterium *Saprospira* Sp. shows colored colonies of shades of yellow, orange, pink and red [70]. The xanthophyll saproxanthin (a modified 3,1'-dihydroxy-γ-carotene) [71] is the major carotenoid. Pigmentation in *Saprospira thermalis* is stimulated by light, by suboptimal concentrations of cobalamin or tyrosine, and by subinhibitory concentrations of thiamine or sodium thiosulfate [70].

Carotenoids are found in the gram-positive lineage, which is separated into two groups according to high or low GC content of genomic DNA [16]. Carotenoids occur frequently in the high GC gram-positive genera. The high GC carotenogenic organisms include carotenogenic actinomycetes, micrococci, nocardias, corynebacteria, and mycobacteria. In the actinomycete lineage, the mycobacteria produce β-carotene, glycosidic carotenoids (phleioxanthophyll) as well as the aromatic carotenoid isorenieratene. *Streptomyces mediolani* also produces isorenieratene and 3 and 3' hydroxyderivatives of the aromatic pigment. Mycobacteria have been reported to produce industrially significant xanthophylls including astaxanthin [72]. Among the corynebacteria, C_{45} and C_{50} pigments and glycosides are present [73]. *Nocardia* sp. contains γ-carotene, 4-keto-γ-carotene, and β-carotene [54], while *Nocardia kirovani* produces xanthophyll glycoside esters. Carotenoid pigments are common in *Micrococcus* sp. The primary pigments are the C_{50} carotenoids sarcinaxanthin [74] and its mono- and di-glycosides [75, 76]. *Micrococcus roseus* contains β-carotene, echinenone, and canthaxanthin [54]. *Brevibacterium* sp. strain KY-4313 produces canthaxanthin as its primary carotenoid [77]. Strain 103 was reported to produce astaxanthin [78].

Of the low GC gram-positive genera, *Staphylococcus aureus* produces C_{30} apocarotenoids [79]. These include the C_{30} analogues of phytoene, phytofluene, 7, 8, 11, 12-tetrahyrolycopene, and neurosporene. The primary pigment is 4,4' diaponeurospren-4-oic acid [80]. *Enterococcus faecium* produces several C_{30} carotenoids similar to those in the staphylococci. The major carotenoid in 15-cis-4-4'-diapophytoene [81] and also present are the xanthophylls diaponeurosporen-4-al, 4,4'-diapolycopenal-4-al, and the glucoside ester of 4,4'-diaponeurosporene [82]. An alkalophilic *Bacillus* sp. contains β-carotene and a C_{50} carotenoid tentatively identified as tetra-anhydrobacterioruberin [83]. Other bacilli apparently also produce carotenoids but they have not been characterized. The motile, marine bacterium *Planococcus* is yellow-orange is color due to carotenoid pigments which consist of unidentified carotenes and xanthophylls [84]. The pigment levels increase as the cells age and as the salt level in the medium increases.

Deinococcus is an unusual genus of gram-positive bacteria that is highly resistant to ultraviolet radiation and heat which may be related to the presence of carotenoids. The structures of the carotenoids have not been adequately characterized but appear to include β-carotene as well as xanthophylls [85]. Sixty percent of the membrane lipids of the related species, *Thermus aquaticus*, a gram-negative thermophile, were reported to consist of carotenoids [86]. Carotenes appear in *Thermus* but the structures apparently have not been reported [86].

Overall, the bacteria produce several unusual pigments including C_{30}, C_{45}, and C_{50} pigments, as well as esters and acyl derivatives. They generally lack typical plant and algal carotenoids such as α-carotene, lutein, and epoxides.

4.2.2 Algae (Protista)

The composition and distribution of carotenoids in the algae has recently been extensively reviewed [87–89] and we will only briefly consider the evolution of the carotenoids in the algae. Pigments characteristic of the various algal classes are listed in Table 2.

The algae (Protista) are not a natural assemblage of organisms but rather represent a highly diverse group of protists [90]. The diversity dwarfs that of the Plantae, Fungi, and Animalia combined [18]. Each major algal class can be distinguished in part by the properties of the chloroplasts, especially by the membrane structure and composition of photopigments [8, 10, 87]. The eukaryotic algae produce unique xanthophylls which are often class-specific [88, 89, 91]. The pigments are usually associated with chlorophylls within plastids but in overproducers extraplastidic carotenoids which do not function as accessory pigments also occur. Extraplastidic carotenoids accumulate in the commercially important green algae *Haematococcus* and *Dunaliella*.

Carotenoids in algae originated from endosymbionts acquired during evolution [92]. The different classes of algae acquired chloroplasts from symbiotic prokaryotes or symbiotic eukaryotic algae. The red algae acquired chloroplasts from symbiotic organisms resembling cyanobacteria [93]. The origins of chloroplasts in other phyla of algae probably evolved from endosymbiotic eukaryotic algae [90, 92]. Dinoflagellates probably acquired chloroplasts by endosymbiososis with a photosynthetic prokaryote. Dinoflagellate chloroplasts typically contain peridinin as the characteristic carotenoid. However, some dinoflagelates also contains unusual carotenoids: *Gyrodinium aureolum* lacks peridinin but contain 19′-hexanoyloxyfucoxanthin as the main carotenoid [94], suggesting possible multiple endosybiotic relationships during its evolution. The promiscuous retention and disposal of plastids by algae and protozoa [92] has probably contributed to the unpredictable array of carotenoids in protists. The sporadic occurrence, loss, and reappearance argues against the use of carotenoids in chemotaxonomy for designating major phyla, a conclusion also supported by other authors [87, 94, 95]. However, carotenoids could be used as tracers to determine the sources of endosymbionts [96]. Since several genes and protein products are involved in carotenoid biosynthesis, Liaaen-Jensen and colleagues have suggested that descriptions of the pathway reactions would be superior to identification of carotenoids for phylogenetic distinctions [87, 97].

The Chlorophyceae is the class of algae of primary industrial interest. They generally accumulate carotenoids similar to those of higher plants, namely α-carotene, β-carotene, lutein, violoxanthin, and neoxanthin but some species also accumulate astaxanthin. The red appearance of some green algae is due to the appearance of extraplastidic carotenoids (secondary carotenoids), often in quite high amounts. Secondary carotenoids usually occur under nutritional (nitrogen or phosphate) or environmental stress (temperature, oxidative, salt, light) conditions in algae living in extreme conditions. Patches of "red snow" are caused by cryophilic chlorophytes, which accumulate red pigments in oil droplets [98] or in the cell membrane. Only a few species in the Chlamydomonadaceae are red-pigmented in the vegetative state [99]. The majority are snow algae or are found in locations subjected to high light intensities. Astaxanthin occurs as a secondary carotenoid in *Chlamydomonas* [100] and *Haematococcus lacustris* [101], and may function as a photoprotectant or "sunshade" [102, 103]. Similarly, massive accumulation of β-carotene in *Dunaliella* has been postulated to protect the alga against excess irradiation [104].

4.2.3 Plants

Terrestrial plants have a limited and uniform complement of pigments consisting of α-and β-carotenes and the corresponding xanthophylls with the hydroxyl groups in the 3 and 3' positions (zeaxanthin and lutein). β-carotene and lutein are the main components. Also occurring in higher plants are the 5,6-epoxides of β-carotene and lutein, violaxanthin and neoxanthin. Other minor carotenoids present are β-cryptoxanthin and antheraxanthin [88, 89]. Carotenoids in plants are located in chloroplasts/chromoplasts where they are noncovalently bound in protein complexes. Some ketocarotenoids such as rhodoxanthin in gymnosperms and cycads, are not present in chloroplasts but in extraplastidic oil droplets [89].

Chloroplasts in terrestrial plants were acquired by a stable endsymbiosis of cyanobacteria [105–107], but during evolution considerable modification of the chloroplast genome has occurred through transfer of genes to the host nucleus and the loss of nonessential genes from the chloroplast. The relatively simple complement of plant carotenoids suggests that a monophyletic stable symbiosis was established that performed several important functions for the plant and became indispensable. This is in contrast to the polyphyletic acquisition and loss of chloroplasts in certain groups of protists [92].

The monophyletic origin of chloroplasts in plants is supported by the identification of two groups of genes that are clustered in chloroplast genomes [108]. DNA sequencing of the chloroplast genome of the red alga *Porphyra purpurea* demonstrated 125 genes, of which 58 ($\sim 46\%$) were not found on the chloroplast genomes of land plants. *crtE* (encoding prephytoene pyrophosphate dehydrogenase) was present in the red algal chloroplast genome but not in the chloroplast of land plants [108]. In land plants, apparently this gene has been transferred to the nucleus. *crtE* is also incoded in cyanelle DNA in the flagellated alga *Cyanophora paradoxa* [109]. In plants, all chloroplast regulatory genes including those for carotenoid synthesis appear to be encoded in the nucleus [108].

4.2.4 Fungi

Recent phylogenetic studies have led to the surprising conclusion that fungi and animals share an evolutionary history which is independent from plants [110, 111]. Sequences of 25 proteins and comparisons of small subunit ribosomal RNA sequences indicated that animals and fungi are a monophyletic group while plants constitute an independent evolutionary lineage [110]. The results imply that animals and fungi share many similarities at the molecular and cellular levels and that yeast or other fungi may be the model system of choice for understanding metabolic processes related to carotenoids in eukaryotes [110].

Carotenoids occur only sporadically in fungi. Of more than 330 species, about 200 species of carotenogenic fungi have been described [89]. The lower fungi and ascomycetes produce mostly carotenes, especially γ-carotene and β-carotene (Table 2). However, some of the brilliantly colored Discomycetes and Pyrenomycetes synthesize carotenoids of increased complexity including phillipsiaxanthin, aleuriaxanthin, and plectaniaxanthin [96]. The richness of the carotenoid diversity increases in the most advanced fungal phylum, the basidiomycetes.

Goodwin [89] proposed several generalizations from his analysis of fungal carotenoids. Among those pertinent to this review are: (1) many fungi produce only carotenes, including the Phycomycetes, Chytridiales, and Blastocladiales; (2) torulene appears in red yeasts and in certain ascomycetes but rarely in Basidiomycetes and never in Phycomycetes; (3) no carotenoids with α-ionone rings (e.g. α-carotene) have yet been isolated, and (4), no xanthophylls characteristic of the green tissues of higher plants and algae (e.g. cryptoxanthin), 5,6-epoxides (except as possible degradation products), or allenes (e.g. neoxanthin) have been demonstrated in fungi. Characteristic fungal xanthophylls are carboxylic acid apocarotenoids such as torularhodin; somewhat similar xanthophylls are found in some fruits (e.g. lycoxanthin) and algae (loroxanthin). Ketocarotenoids including echinenone, canthaxanthin, and astaxanthin are also occasionally found in fungi [89]. Carotenoids typically found in the four subdivisions of fungi (Zygomycotina, Ascomycotina, Basidomycotina, and Deuteromycotina) are listed in Table 2.

Carotenoids are relatively rare in yeasts confined only to genera of heterobasidiomycetes [112]. They unfortunately do not occur naturally in common industrial yeasts such as *Saccharomyces cerevisiae*, *Kluyveromyces*, and *Candida utilis*. Carotenoids characteristic of the red heterobasidiomycetous yeasts are torulene and the carboxylic acid torularhodin, which appear *Rhodotorula*, its perfect stage *Rhodosporidium*, in *Sporobolomyces*, *Sporidiobolus* and in some species of *Cryptococcus*. Plectaniaxanthin, a pigment normally present in Discomycetes, was detected in the imperfect yeast *Cryptococcus laurentii* [113]. The presence of carotenoids is used as a criterion in yeast classification [112]. This appears reasonable since carotenoid synthesis involves several genes and is expressed in only a limited group of yeasts although it can be an unstable property.

The discovery of *Phaffia* in the 1960's [114], the description in 1976 of its unusual physiological characteristics compared to other pigmented yeasts [114,115], and the elucidation of the structure of its pigments [116,117] has added important diversity to yeast carotenoids. *Phaffia* produces astaxanthin, 3-hydroxy-3',4'-didehydro-β,ψ-caroten-4-one (HDCO), torulene and β- and γ-carotene as its primary pigments. The levels of pigments in wild isolates are low but they have been increased greatly by strain development in industry. *Phaffia* accumulates the 3R and 3R' configurational isomers which are the opposite configuration found in *Haematococcus*:

Certain fungi have the ability to accumulate cartenoids to quite high levels. For example, *Phycomyces* can accumulate up to 25 mg of β-carotene per g of mycelium when cultivated on agar plates and subjected to genetic activation and external agents [118]. A high capacity to synthesize carotenoids is evident another fungi as well, e.g. cells of some jelly fungi (*Dacrymyces*) are packed with oil globules containing carotenes. Carotenoids in fungi probably accumulate in lipid globules and in membranes, and may provide antioxidant protection.

The evolutionary origins of carotenoids in fungi are not known. The choanoflagellates or perhaps the true slime molds may be the closest protistan ancestor to the fungi [119]. However, like the algae, it is probable that the source of endosymbionts in fungi is polyphyletic. There is strong homology between the bacterial and fungal genes encoding enzymes early in the isoprenoid and carotenoid pathways [120], but these genes may be universally congruent in all carotenogenic organisms. Fungi may have acquired the ability to produce carotenoids by endoymbiosis of bacteria, algal (e.g. Rhodophyta or Chlorophyta) chloroplasts, or by horizontal gene transfer from bacteria. Intimate pairing in lichens may also have encouraged symbiotic transfer of carotenogenic genes. The finding of phillipsiaxanthin and plectaniaxanthin in fungi, which share structural characteristics with photosynthetic bacteria, support the polyphyletic nature of fungal carotenoids. If fungi acquired plastids as a source of carotenoids, it is possible that a morphologic or genetic study would reveal some relics of this plant and algal organelle in the carotenogenic fungi. Alternatively, the findings that fungi and animals are sister groups indicates a common ancestor which may have been an unidentified protist. Suggestions for the ancestor have included choanoflagellates [119, 121] or certain slime molds. The choanoflagellates lack carotenoids [122] whereas many slime molds are pigmented. The slime mold *Lycogola epidendeon* in the class Myxomycetes produces γ- and β-carotenes, 3,4-didehydrolycopene, torulene, and neurosporaxanthin [113].

The lichens, which are symbiotic associations between fungi (often ascomycetes) and green algae or cyanobacteria, contain a wide diversity of carotenoids. This is expected considering the polyphyletic origins of carotenoids in algae and fungi. Extensive studies of carotenoid composition in lichens have been carried out by Czeczuga and his colleagues in Poland and recently summarized [123].

The carotenoid composition of lichens depends on their exposure to light. Both typical algal and fungal carotenoids appear in lichens. Lutein, epoxides, α-carotene, zeaxanthin and other typical plant and algal carotenoids have been reported. Fungal carotenoids including torulene are also present. Several carotenoids of industrial interest occur in lichens including β-carotene, zeaxanthin, canthaxanthin, lutein, and astaxanthin. Most of the carotenoids in lichens are present in low concentrations ($\leq 100 \, \mu g \, g^{-1}$). Nonetheless the lichens could be valuable as biological material for isolation of new genetic sources of carotenoids and also may be of interest because of possible genetic and physiologic interchange between the symbiotic pairing of fungi and algae.

5 Biological Functions of Carotenoids

Interest in carotenoids has increased considerably during the past ten years, due in part to the growing evidence of benefits to human health and also to the growth of certain areas of agriculture, especially aquaculture and the poultry industry. In the field of human benefits of carotenoids, several proposed attributes and protective roles in chronic disease are speculative, but clinical trials should critically evaluate the benefits during the next few years. As roles in human disease prevention become known, the commercial demand for carotenoids should increase.

5.1 Roles of Carotenoids in Preventing Degenerative Diseases in Humans

The possibility that carotenoids have roles in preventing cancer and other degenerative diseases has generated considerable interest and activity within the carotenoid field. The primary degenerative diseases associated with aging are cancer, cardiovascular disease, immune-system decline, brain dysfunction, arthritis, and cataracts [124]. Epidemiological investigations have shown that consumption of green and yellow vegetables is associated with reduced incidence of some cancers [125, 126]. In fact, good dietary consumption of fruits and vegetables approximately halves the likelihood of developing cancer [124]. Research has supported the theory that life expectancy is related to oxidative damage of DNA, proteins, and lipids [124, 127 128] and that natural antioxidants including carotenoids among other substances are important in the prevention of these oxidations.

An important factor in longevity appears to be the basal metabolic rate [124] since metabolism produces oxidative by-products which can cause extensive damage to cellular components. The primary oxidants generated in metabolism

are superoxide (O_2-, hydrogen peroxide (H_2O_2), hydroxyl radical, lipid epoxides, hydroperoxides, lipid alkoxyl and hydroperoxides, and singlet oxygen $^1O_2$.
All of these are products of normal metabolism and millions of each species are
produced each day [124]. Singlet oxygen is formed from transfer of energy from
light, the respiratory burst of neutrophils, lipid peroxidation, or other dark
reactions [129–131]. Singlet oxygen has benefits as well as detriments: it acts as
a meditator of the antibacterial action of leukocytes [132] but its reactivity with
human tissues can cause damage and it probably contributes to acute and
chronic health disorders [133–138]. Antioxidant defenses against these damaging agents include ascorbate, tocopherol, and carotenoids [124]. One of the first
classes of compounds demonstrated to effectively quench $^1O_2$ was the carotenoids [139]. Carotenoids physically quench $^1O_2$, generating ground state
oxygen and a carotenoid triplet which returns to its ground state with release of
heat [140]. Carotenoids vary in their quenching capacity; $^1O_2$ quenching
capacity generally increases as the number of conjugated double bonds in the
carotenoid increases, but quenching also varies with chain structure, functional
groups, *cis/trans* isomerism and the technique used for determination
[141, 142, 143].

β-Carotene has been proposed as an important dietary anticarcinogen, and
evidence has been obtained that β-carotene prevents cancers caused by chemicals and viruses [144–146]. However, several other carotenoids occur in plant
foodstuffs including α-carotene, γ-carotene, lycopene, lutein, neoxanthin, violaxanthin, and others. Murakoshi et al. [147] presented evidence that α-carotene
was more effective than β-carotene in preventing lung and skin carcinogenesis in
mice. Recently, it was shown that astaxanthin but not canthaxanthin effectively
prevented urinary bladder carcinogenesis [148]. Since astaxanthin is not a precursor to vitamin A in animals, the action of astaxanthin may have been related
to suppression of cell proliferation. Astaxanthin was shown in vitro to enhance
antibody production to sheep red blood cells in normal mice [149, 150]. This
immunomodulating activity could reduce the chance of developing autoimmun-
ity and malignancies by enhancing T-helper functions and promoting specific

Table 3. Functions and uses of carotenoids in humans–known and proposed [153, 154]

Quenching of singlet oxygen
Termination of lipid peroxidation
Cancer prevention
Immunomodulation
Gap junction cell-cell communications
Modulation of lipoxygenase activity
Cardiovascular disease prevention
Cataract prevention
Photosensitivity prevention
Vitamin A precursor-vision, cell differentiation
Membrane modulators, fluidity control
Modulators of membrane enzymes
Precursors of hormones, vitamins, e.g. retinoids for human development

---

antibody responses [149, 150]. Astaxanthin is an excellent quencher of $^1O_2$ compared to several other carotenoids tested, which may contribute to its anticarcinogenic activity. Damage to ocular tissue occurs by radiant energy in the 295–450 nm range [151], and blue light and near-UV radiation damage is often increased by natural photosensitizers in cells (i.e riboflavin). Of the mechanisms by which radiant energy damages ocular tissues, the involvement of $^1O_2$ has been best documented [151, 152]. It is possible that ocular tissue, damage could be lessened by carotenoids. Another example is the formation of oxidized low-density plasma lipoprotein (LDL), which contributes to cardiovascular disease. Carotenoids prevented oxidation of human LDL [153]. Other biological functions of carotenoids in humans have been proposed and are being investigated in animals and in human clinical trials (Table 3) [154, 155].

5.2 Functions of Carotenoids of Microorganisms, Plants, and Animals

Carotenoids and their derivatives are known to have several functions in photosynthetic bacteria and eukaryotes. Besides their universal role in photosynthetic organisms as light-harvesting pigments and as photoprotectants (quenching of triplet-state prophyrins and $^1O_2$) in photosynthetic membranes [1–3], they have several additional biological functions. In corn coleoptiles, zeaxanthin has been proposed as the receptor for blue light-induced phototropism [156]. Carotenoids are a source of hormones and stress response in plants: the growth regulator abscisic acid is formed in water stressed plants by cleavage of carotenoids [157]. In algae, carotenoids are cleaved to retinoids, which combine with opsin to form the photoreceptor for phototaxis [158].

The well-established function of carotenoids in animals is as vitamin A precursors [159] and as a source of retinoids crucial for vertebrate development [160]. In addition, they probably have a role in quenching singlet oxygen and terminating lipid peroxidation [161]. This activity possibly prevents mutations in the ova of marine fish and shellfish which may be a reason the eggs of many of these species are so highly pigmented. Many of the functions proposed above for humans may apply to lower animals as well.

Carotenoids in oceanic animals probably have their origin in phytoplankton at the ocean surface. Still, deep sea animals distant from the surface waters often have substantial concentrations of carotenoids obtained through the food chain. The function of carotenoids is probably for reproduction, osmotic tolerance, or other purposes, and not for photoprotection since these animals are not exposed to bright light. In contrast, hydrothermal vent animals generally lack carotenoid pigmentation, which is consistent with the nutritional input of these animals coming from carbon fixed within the vent area and not deriving from photosynthetically fixed carbon at the ocean surface. The exceptions are the vent decapod crustaceans [162]. Although the exoskeletons of deep-vent shrimp and crabs are nearly colorless, their eggs and hepatopancreas are colored, though less carotenoid is present than in benthic crustacea [162]. The carotenoids present in the vent brachyuran crab *Bythograea thermhydron* include astaxanthin (80%), β-carotene (10%) and the remainder echinenone, canthaxanthin and presumably carotenes. Interestingly, α-carotene was not detected suggesting that the carotenoids were not of algal origin. The restricted composition of pigments suggests that the pigments may be synthesized by bacteria in the thermal vent environment which are then ingested. The vent habitat could offer a new location for isolation of novel organisms producing astaxanthin and other pigments.

Carotenoids have several roles in fungi. They are the precursors to mating hormones (trisporic acids) in the Mucorales. They may also be involved in development and reproduction in *Neurospora* since biosynthesis of carotenoids is photoregulated in the mycelium but not in the conidia. The most general function is to protect against $^1O_2$ and phytoalexin/oxygen mediated killing of fungi in the phyllosphere. Carotenoids can stabilize membranes and lessen damage by microbiocidal medium chain fatty acids common on the leaf surface. Cross-linked carotenoids in pollens and in the zygospores of ascospores of some fungi provide extraordinary resistance to chemical and environmental stresses [163]. They may also have antimicrobial properties in certain ecological niches: the pepper carotenoid capsanthin inhibits growth and aflatoxin production by *Aspergillus* [164].

Carotenoids can stabilize lipid membranes by decreasing water permeability and increasing rigidity [165–167]. Carotenoids decreased membrane fluidity and increased resistance to oleic acid killing in *Staphylococcus aureus* [168]. This decrease in membrane fluidity could also contribute to heat and radiation damage. The resistance provided to organisms by carotenoids may have ecological significance allowing organisms to colonize certain environments, for example staphylococcal invasion of wounds, myocbacteria in AIDS, and colonization of leaf surfaces by pigmented heterobasidiomycetous yeasts.

6 Biosynthesis of Carotenoids

Excellent recent reviews of the biosynthesis of isoprenoids to cholesterol and carotenoids have been published [169–171], and only some new developments are presented here, particularly as they pertain to microbial synthesis of industrially important carotenoids.

A crucial aspect of the regulation of biosynthesis of carotenoids in microorganisms is the location of synthesis, the spatial separation of carotenoid biosynthesis from sterol and lipid synthesis (which also use acetyl-CoA as a substrate), and the deposition and transport of isoprenoids throughout the cell. The rate limiting enzyme in the isoprenoid biosynthetic pathway in *Saccharomyces cerevisiae* and in mammalian cells is 3-hydroxy-3-methylglutaryl-coenzyme

A reductase (NADPH) (i.e. HMG-CoA reductase). In contrast to mammalian cells which have a single reductase, the yeast has two functional genes encoding the enzyme with limited homology to HMG-CoA reductase from hamster cells [172]. Plants also contain multiple reductase genes which are differentially regulated. *Arabidopsis thaliana* contains two differentially expressed reductase genes, whose product is inserted into the endoplasmic reticulum-derived microsomal membranes [173]. One of the yeast reductases (Hmg1p) resides in the perinuclear ER, while the other is present in peripheral ER membrane [174]. The different cellular locations may allow specific compartmentalization of mevalonate synthesis. One of the two isozymes (Hmg2p) is rapidly degraded in response to feedback signals from the isoprenoid pathway [175]. Decreased mevalonate pathway flux reduces the rate of degradation of Hmg2p, and one signal is a nonsterol intermediate prior to squalene. Overexpression (10-fold) of Hmg1p causes the proliferation of the membrane in which the reductase resides, forming perinuclear stacked structures termed karmellae and thus increased compartmentalization of the enzymes [176]. Selective protein degradation in the ER is controlled partly by redox potential [177], which could be influenced by respiratory activity of the cells or cyanide-insensitive oxygen uptake by cytochrome P-450s which are abundant in the ER.

Sterol synthesis in *S. cerevisiae* is differentially regulated and partitioned in different compartments [178]. The HMGR isozymes directed essentially equal quantities of carbon to the biosynthesis of sterols when heme was available despite a large (67-fold) difference in specific activity of the isozymes. Palmitoleic acid (16:1) acted as a positive regulator and ergosterol and oleic acid (18:1) as inhibitors but the regulation pattern differed for each isozyme. Casey et al. [178] concluded that separate compartmentalized isoprenoid pathways exist in *S. cerevisiae*. Carotene and sterol pathways in *Phycomyces* were reported to be independently regulated and physically separated in different subcellular compartments [179]. The reactions of carotene biosynthesis in *Phycomyces* and zeaxanthin synthesis in a *Flavobacterium* sp. were also tightly channeled in the cell because of the presence of enzyme aggregates [180–183]. Enzyme complexes have also been observed for the reactions in early isoprenoid synthesis leading to phytoene [184]. Therefore, cellular topography, compartmentation, and the presence of enzyme aggregates adds a hierarchy of complexity to carotenoid biosynthesis, and conclusions from in vitro 'solubilized' enzymes on carotenoid biosynthesis will have to be made cautiously.

By analogy with other isoprenoid systems, carotenes probably originate in the smooth ER and then are transported within the cytoplasm in lipid vesicles or in complexes with lipoproteins. Transmission electron micrographs of sections of *Phaffia rhodzyma* and comparison to confocal autofluorescent images indicates that carotenoids are deposited in lipid globules, which in wild-type yeasts are mainly associated with the lipid membrane and are especially prevalent in the growing tip of buds. In hyperproducing mutants the fluorescence was more widely distributed throughout the cell. Xanthophylls may also be biosynthesized in the ER since cytochrome P-450 and cyanide-insensitive oxygen uptake is abundant in this organelle. Alternatively, carotenes could be formed in the ER, and deposited in lipid globules or ER vesicles, and transported to other membraneous regions or remain in the cytosol depending on the strain of yeast and its rate of production. Once in the membranous region, the carotenes may then react enzymatically with oxygen metabolites to form the characteristic xanthophylls depending on the species. Secondary carotenoids (astaxanthin and other ketocarotenoids) in *Haematococcus* first accumulated in lipid bodies surrounding the nucleus [185], and were suggested to function as a physicochemical barrier protecting the genome from free-radical damage [186]. The smooth ER in yeasts is not only associated with the nuclear region but also may be contiguous with the cytoplasmic membrane and growing tips of the bud. Membrane-bound, small, ER derived vesicles at the growing tips of the bud have been observed in *S. cerevisiae* and in the heterobasidiomycetous yeasts *Rhodosporidium* sp. and *Sporidiobolus salmonicolor* [187]. The quantity of ER in yeast cells varies greatly, and is related to growth rate and oxygen availability [188], conditions which also affect carotenoid synthesis.

The ontogeny of lipid bodies in plants and *Neurospora crassa* was examined by electron microscopy during massive lipid accumulation [189]. The lipid bodies as well as the outer membrane of plastids of fat-storing cells were associated with ER-like cytoplasmic membrane structures, suggesting that both plastids and ER may be involved in the synthesis of lipid bodies. Chromoplast development in fruits and flowers of plants results during differentiation of plastids and is often associated with accumulation of large quantities of carotenoids [190]. Carotene bodies contribute to the morphology of the mature chromoplast. Depending on the type of chromoplast, carotene molecules crystallize, dissolve in lipid globules, or become tightly bound to protein molecules. More than 60% of the plastid membrane protein participate in carotene binding or synthesis. The ontogeny of the chromoplasts appears to offer an ideal system for investigating the biosynthesis of carotenoids.

It is likely that carotenoids are formed and deposited in the cell in the proximity of enzyme systems that generate toxic forms of oxygen such as mitochondria or microbodies. Microbodies (peroxisomes, glyoxosomes, glyoxiperoxisomes) are frequently associated with strands of endoplasmic reticulum [191]. Carotenoid biosynthesis and modification in eukaryotic cells may occur in proximity to peroxisomes, since these organelles are involved in oxidative reactions and generate and consume H_2O_2. Similar to the mitochondrion, the peroxisome is a major site of oxygen utilization, and may represent the vestige of an ancient organelle that carried out all of the oxygen metabolism in procaryotic or primitive eucaryotic cells. The protein composition of peroxisomes is limited and the total peroxisomal protein consists of up to 40% catalase. Peroxisomes are responsible for the degradation of alcohol to acetaldehyde and for the degradation of fatty acids to acetyl-CoA producing H_2O_2 in the process. The acetyl-CoA generated from fatty acid breakdown can be transported to the mitochondria for utilization in the citric acid cycle, or it could be used for biosynthetic reactions such as isoprenoid biosynthesis. The metabolic roles of peroxisomes have been investigated in green leaves. At high light intensities and in hypoxic conditions, chloroplasts generate the 2-carbon compound glycolic acid, which exits the chloroplasts and enters into the peroxisomes. Glycolic acid is oxidized forming hydrogen peroxide and glyoxylate, and glyoxylate can then be converted to the amino acid glycine for protein synthesis or transported to mitochondria for further metabolism. Oxidized glycolic acid can also condense with two molecules of acetyl-CoA from fat degradation forming isocitric acid, and degraded to succinate and glyoxylate, thus perpetuating the cycle. Populations of peroxisomes and mitochondria increased in population during leaf senescence and were associated with the transition of leaf peroxisomes into glyoxysomes [192]. Aging and senescence in fungi is associated with increased generation of reactive oxygen metabolites and differentiated cells have a relatively more prooxidizing or less reducing intracellular environment [193].

The involvement of peroxisomes in isoprenoid biosynthesis has been demonstrated in mammalian cells. Mevalonate kinase has been localized to rat liver peroxisomes [194]. Isoprenoid biosynthesis or squalene has been demonstrated in mammalian cells [195]. The physiological importance of peroxisomal isoprenoid formation to overall polyisoprenoid synthesis is not yet established. The synthesis in peroxisome could contribute to compartmental regulation of isoprenoid synthesis.

7 Genetics of Carotenoids

The genetics of carotenoid biosynthesis in bacteria, plants, and fungi has been recently reviewed [169, 170] and only selected areas are discussed. Genes for carotenoid biosynthesis, i.e. the *crt* genes, are clustered on the chromosome in many eubacteria such as *Rhodobacter capsulatus* [196], *Rhodobacter sphaeroides* [197], *Erwinia herbicola* [198, 199], *Erwinia uredovora* [200], and *Mycobacterium aurum* [202]. The structures are unlinked to the regulatory genes for carotenogenesis in *Myxococcus xanthus* [201]. A carotenogenic gene cluster containing the *crtB* gene is on a large plasmid in the gram-negative thermophile *Thermus thermophilus* [203]. Gantotti and Beer [204] reported that pigmentation and thiamine prototrophy in *Erwinia herbicola* was controlled by a large plasmid. The genes controlling conversion of lycopene into xanthophylls were not linked to the gene cluster encoding carotene biosynthesis in *M. aurum* [202]. These reports indicate that differences exist in the genomic locations and expression of *crt* genes in certain nonphotosynthetic and photosynthetic bacteria. The carotenoid biosynthetic genes in purple nonsulfur bacteria are present in superoperons, in which transcription is coupled to genes encoding bacteriochlorophyll [205]. This arrangement enables rapid expression and assembly of the photosynthetic apparatus when oxygen concentrations become limiting.

The *crt* genes encoding carotenoid biosynthesis in *E. herbicola, E. uredovora* and *R. capsulatus* have been expressed in *E. coli* [200, 206, 207]. The *crt* genes of *R. sphaeroides* were expressed in nonphotosynthetic bacteria including *Paracoccus denitrificans, Agrobacterium tumefaciens,* and *Azotomonas insolita* [208], which is interesting from an evolutionary point of view because it indicates that nonphotosynthetic bacteria could acquire the ability to synthesize carotenoids from photosynthetics through a symbiotic relation or by genetic exchange of the carotenogenic genes.

Among carotenogenic organisms, the *crt* and regulatory genes in the purple nonsulphur photosynthetic bacteria *Rhodobacter capsulatus* and *Rhodobacter sphaeroides* and in the chemoorganotrophs *Erwinia herbicola* and *Erwinia uredovora* have been studied most extensively. *Rhodobacter* synthesizes acyclic carotenes and methoxy carotenoids while *Erwinia* forms cyclic and hydroxylated pigments including β-carotene and zeaxanthin and glycosides of zeaxanthin. Eight carotenogenic genes of *R. capsulatus* are organized in 4 operons: *crtA, crtIBK, crtDE* and *crtEF*. The ninth gene, *crtJ*, lies on a fragment separated by 12 kb. Many of the *R. capsulatus* gene products have been identified. *crtE* encodes phytoene synthetase, *crtB* and *crtJ* code for prephytoene pyrophosphate synthetase, while *crtI* has been found to code for phytoene dehydrogenase. However, in *Thermus thermophilus, crtB* was found to code for phytoene synthase. By introducing *crtB* on a multicopy plasmid into a hyperproducing mutant, carotenogenesis was increased 20 fold. This suggests that phytoene synthase is a rate limiting step for carotenogenesis in this organism [209]

In *Erwinia* the carotenogenic genes coding for the first three enzymes in the pathway (prephytoene synthase, phytoene synthase and phytoene dehydrogenase) share a common nomenclature with *Rhodobacter* (*crtB, crtE* and *crtI*). The remaining three genes, however, coding for lycopene cyclase, β-carotene hydroxylase and zeaxanthin glycosylase have different nomenclature in the two *Erwinia* species. In *E. herbicola* they are *crtZ, crtH* and *crtG* respectively, while in *E. uredovora* they are *crtY, crtZ* and *crtX*. Although the GC contents of *Erwinia* and *R. capsulatus* are different (52.6–57.7% in *E. herbicola,* 53–54.5% in *E. uredovora* and 65.5–66.8% in *R. capsulatus*), and the organisms are not closely related by several other properties, the sequences of the putative proteins encoded by *crtI, crtB* and *crtE* all share homology [210]. *crtB* and *crtE* gene products also have similar domains among various prenyltransferases involved in isoprenoid biosynthesis [210]. In *Rhodobacter, Erwinia,* and *Neurospora,* a single gene encodes a gene product (CrtI/A1–1) which performs multiple dehydrogenations to neurosporene or even lycopene, while cyanobacteria, green algae, and higher plants presumably require two gene products [211]. The carotenoid dehydrogenases of *R. capsulatus* and *E. herbicola* are homologous (26 to 76%) and contain an evolutionarily conserved FAD/NAD(P) and carotene binding regions, while the dehydrogenase from *Neurospora crassa* is more distantly related. Phytoene dehydrogenase in tomato, a single polypeptide catalyzing the conversion of phytoene to ξ-carotene, is transcriptionally regulated during fruit ripening [212]. The deduced amino acid sequence of the tomato desaturase shows high homology to phytoene desaturases of cyanobacteria and algae, but was unrelated to those from purple bacteria and fungi. Thus, two evolutionarily unrelated lineages of phytoene desaturases appear to exist in nature.

The genetics of fungal carotenoids have mostly been investigated in *Phycomyces* and *Neurospora crassa*. Carotenoid pigments are induced by blue light in *N. crassa*, and carotenogenesis is also induced developmentally in the dark during formation of macroconidia [213]. Colorless (albino) mutants are common in *N. crassa*, in which *al*-1 mutants accumulate phytoene [214]. Four genes of *N. crassa* encode enzymes for carotenoid biosynthesis, *al-1*, *al-2*, *al-3*, and the yellow-1 locus [214]. Mutations in any of the three albino genes gives white mycelia and conidia. The *al-1*, *al-2*, and *al-3* genes encoding phytoene dehydrogenase, phytoene synthase, and geranylgeranyl pyrophosphate synthetase have been cloned and sequenced [215–217]. All three genes are photoinduced by blue light. Phytoene synthase activity is membrane associated in *Neurospora* [218]. The phytoene dehydrogenase deduced amino acid sequence has homology to those of *Rhodobacter* and *Glycine max* (soybean). Bacterial and eukaryotic phytoene synthases and prephytoene pyrophosphate synthase proteins share homology [219], supporting the hypothesis of structural relatedness of very early carotenoid biosynthetic enzymes (phytoene synthases and prephytoene pyrophosphate) in phylogenetically diverse organisms, while two distinct lineages of phytoene desaturases apparently have evolved. Further divergence is expected for later enzymes, but the genes and enzymes catalyzing the formation of xanthophylls in different organisms remain to be isolated and characterized.

Carotenoid expression in various bacteria and fungi is controlled by several mechanisms including blue light, oxygen, conidial development, and growth stage. In purple nonsulfur bacteria (*Rhodobacter*), the "superoperons" harboring *bch* and *crt* genes, appear to be required for optimal transition of cells from aerobic respiration to anaerobic photosynthesis. Genes encoding the superoperons are repressed by a mechanism involving oxygen. In low oxygen conditions, the expression of the photosynthetic apparatus is regulated by light intensity [220, 221]. Little is known of the mechanism governing the oxygen-mediated expression of the *crt* operons [222]. A gene (*ppsR*) was isolated from the photosynthetic gene cluster of *R. sphaeroides* that represses Crt protein levels [222]. Inactivation of the *ppsR* gene causes overproduction (5-fold) of carotenoid pigments and bacteriochlophylls, but these mutants are unstable [222]. The deduced amino acid sequence of the protein product of *ppsR* is homologous to the CrtJ protein of *R. capsulatus*, and also to the C-terminal region of various other regulatory proteins, which have similar structure as response regulators of two-component regulatory systems. While ppsR appears to be a transcriptional repressor, the mechanism by which oxygen affects expression is not known. Oxygen or a derivative could react with transcriptional regulators directly leading to an oxidized inactive state, or oxygen could be sensed at the cell membrane by a redox active two component system.

Other bacteria apparently regulate carotenogenesis by other mechanisms. In *Myxococcus xanthus*, carotenoids accumulate after the transition from exponential to stationary phase and their formation is absolutely dependent on light [224]. Expression of carotenoids in stationary phase is due to exhaustion of the carbon source [224], a condition that also stimulates the cell's response to oxidative stress [225]. Hodgson and Murillo [224] have proposed the interesting hypothesis that $^1O_2$ is involved directly in activation of the *carORS* operon in *M. xanthus*. Carotenoid formation in *M. xanthus* seems to depend on protoporphyrin IX as a photosensitizer [226], which has an absorption maximum of 410 nm. In a model in which expression is regulated by $^1O_2$, blue light excites protoporphyrin IX in the membrane to the triple-state, which in turn reacts with O_2 to form $^1O_2$. Reaction of the membrane bound CarR causes release of CarQ and expression of genes encoding carotenogenic enzymes as well as CarR. The carotenoids in turn quench $^1O_2$ and shut down expression.

The genetics of β-carotene hyperproduction in *Phycomyces* has been investigated in depth by Cerda-Olmedo and his coworkers. Carotene and sterol synthesis in *Phycomyces* are physically separated in different compartments and the biosynthetic reactions of carotene biosynthesis are channeled because of an enzyme aggregate [227]. β-Carotene production in *Phycomyces* is dramatically affected by culture conditions, the strain, chemical stimulators, illumination, sexual activity, and endogenous genetic changes. Mutations in the genes *carS*, *carD*, and *carF* dramatically increased carotenoid levels. In contrast, carotene levels were decreased by mutations in *carA* and *carC*. Sexual interactions between heterothallic Mucorales gave heterokaryons and increased the carotene content. Physical contact for sexual stimulation was not necessary, which led to the isolation of diffusible trisporic acids that are produced from β-carotene. A third impressive component of carotene overproduction in *Phycomyces* is chemical activators. Several chemicals added to the medium increase the β-carotene content [228]. One group of chemical activators such as retinol and β-ionone are related in structure to β-carotene, while another group are phenols. When sex is combined with *carS* and *carF* mutations, the carotene content increased from 100 to 25000 μg g^{-1} mycelium [118]. End-product regulation by β-carotene is mediated by the *carS* gene product. Mutations in *carS* result in massive accumulation of the intermediates lycopene and phytoene whereas others were white because the *carS* product stops carotene synthesis with very low levels of β-carotene [229]. The *carS* gene product also combines with the main regulator *carA* product pA, a complex which represses expression of the carotenogenic genes. pA activity is affected by light and sexual activity. The genes of the isoprenoid pathway in *Phycomyces* are also investigated at the sequence level [230].

Despite the growing commercial importance of *Phaffia rhodozyma*, very little is known of its genetics. In part, this was due to the inability to find sexual activity. Quite recently Golubev [231] reported mother-daughter cell conjugation on polyol-containing media. This resulted in holobasidia formation with terminal basidiospores. No discharge of basidiospores, nor mycelium formation were reported. This telomorphic stage was given the name *Xanthophyllomyces dendrorhous.* Golubev [231] suggested that *Phaffia* had been described much earlier by Ludwig [232, 233] and named *Rhodomyces dendrorhous.*

By using Contour-clamped Homogeneous Electrical Field (CHEF) gel electrophoresis of 6 strains of *P. rhodozyma*, it has been shown that heterogeneity exists in chromosome size [234]. While most chromosomes have a size of 1000–2000 kb, some strains were found to have an additional chromosome of 3000 kb. Most, but not all strains contained an additional chromosome of 250–350 kb. The total chromosomal number was 7 in two of 6 strains studied and 6 in the remaining 4 strains. When the chromosomes in the CHEF gels were transferred to nitrocellulose and probed with the *S. cerevisiae ura3* and *S. pombe ura4* genes the *ura3* sequence (but not the *ura4* sequence) hybridized to *Phaffia* genomic DNA. The total DNA content of *P. rhodozyma* has been estimated by flow cytometry to be $\sim 4 \times$ the content of haploid *S. cerevisiae* [235].

Although certain classes of auxotrophic and other mutations in essential genes have been difficult to obtain in *Phaffia* in some laboratories, color mutants occur commonly. Girard et al. [236] generated a series of color mutants from a wild-type strain by EMS and UV mutagenesis. The color of the mutants ranged from white to intense red. Twenty eight color mutants were recovered and carotenoid analysis showed three main classes: white (lacking carotenoids or accumulating phytoene), yellow mutants that accumulated β-carotene, and astaxanthin-hyperproducers. Interestingly, in several β-carotene and phytoene mutants, the levels of these pigments occurred at much higher levels than in the wild-type. This suggests that astaxanthin or another xanthophyll may regulate the pathway by feedback inhibition. Similar results have been observed in our laboratory in experiments where astaxanthin was decreased by chemical treatment [237]. According to the carotenoid analysis by Girard et al. [233], no mutants were obtained that accumulated carotenes at quantities higher than β-carotene. Nor were mutants obtained that accumulated xanthophylls in the Andrewe's sequence between β-carotene and astaxanthin, except for two mutants that produced mostly echinenone and hydroxechinenone and one strain that accumulated an unknown xanthophyll (presumably 3-hydroxy-3',4'-didehydro-β-ψ-caroten-4-one [HDCO]). Only two of the isolated mutants produced higher levels of total cartenoids than the wild-type and both of these grew poorly compared to the parent. One yellow mutant produced nearly 2 mg of β-carotene g^{-1} when yeast extract/peptone medium was supplemented with 3% glycerol.

8 Industrial Aspects of Microbial Carotenoids

8.1 Industrially Important Carotenoids

Although more than 600 carotenoids have been identified from various sources, only about 40 are ingested by humans, and only a handful are used industrially for food colors, animal feeding, pharmaceuticals and cosmetics [238, 239]. The primary colorants used in foods and feeds are presented in Table 4.

Most of the carotenoids used in the food, feed, nutritional supplement, cosmetic, and pharmaceutical industries, and for other industrial purposes are produced by total chemical synthesis. β-Carotene has been produced commercially since 1954 [11]. Six synthetic carotenoids or apo-derivatives have become important commercially [11]: β-apocarotenal, β-apo-8'-carotenoic acid ethyl ester, citranaxanthin (5',6'-dihydro-5'-apo-18'-nor-β-carotene-6'-one), β-carotene, canthaxanthin, and racemic astaxanthin (Fig. 4). According to Pfander [11], approximate current market prices for stabilized dispersible powders containing 5–10% active carotenoid are $600 \, kg^{-1}$ for β-carotene, $900 for β-apo-8'-carotenoids, $1300 for cantaxanthin, and $2500 for astaxanthin. Total sales for synthetic carotenoids was estimated to be about $300 million and may be more than $500 million by 1997 [11]. The world market estimates for synthetic and biological sources of astaxanthin, canthaxanthin, and β-carotene are presented in Table 5 [240]. It is nearly impossible to estimate the market for natural extracts since many of the companies are privately held and in development. Of course, the market and demand for carotenoids may change if medical benefits are demonstrated in clinical trials.

Table 4. Carotenoid sources used as colorants in foods and in feeds

A. Natural extracts

Compound	Major Carotenoids
Annatto (*Bixa orellano*)	Bixin, norbixin
Carrot oil (*Daucus carota*)	α-Carotene, β-carotene
Orange peels	Lutein esters
Palm oil	Carotenes, lutein
Paprika (*Capsicum annuum*)	Capsanthin, β-carotene, cryptoxanthin, capsorubin
Tomato (*Lycopersicon esculentum*) extracts	Lycopene, β-carotene
Saffron (*Crocus sativus*)	Crocin, β-carotene, zeaxanthin
Yellow corn (*Zea mays*)	Lutein, zeaxanthin, cryptoxanthin, carotenes
Alfalfa meal	Lutein, zeaxanthin, cryptoxanthin, violaxanthin, neoxanthin
Marigold extracts (*Tagetes erecta*)	Lutein (ester), β-carotene

B. Culture products
Phaffia rhodozyma	Astaxanthin, HDCO, β-carotene
Blakeslea trispora	β-Carotene
Dunaliella salina	β-Carotene
Haematococcus sp.	Astaxanthin

C. Synthetic carotenoids
β-Carotene
β-Apo-8'-carotenal
β-Apo-8'-carotenoic esters
Canthaxanthin
Citranaxanthin
Astaxanthin

Presently, carotenoids for industrial purposes are nearly exclusively obtained by chemical synthesis or by extraction of plant materials, but these carotenoids also occur naturally in species of microalgae, bacteria, and fungi. Much interest has been devoted to microbial synthesis of carotenoids, and several patents have been filed in recent years (Table 6). Microbial carotenoid products from *Dunaliella*, *Haemotococcus*, and *Phaffia rhodozyma* are now being commercialized.

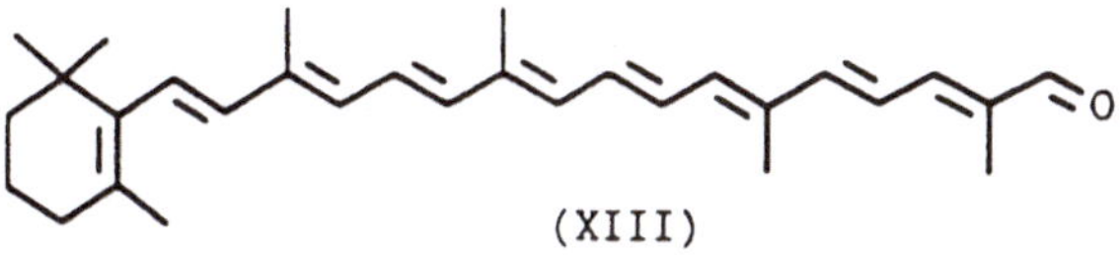

(XIII)

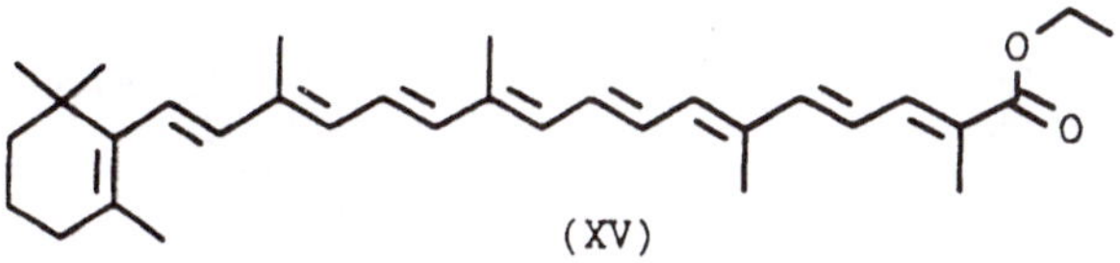

(XV)

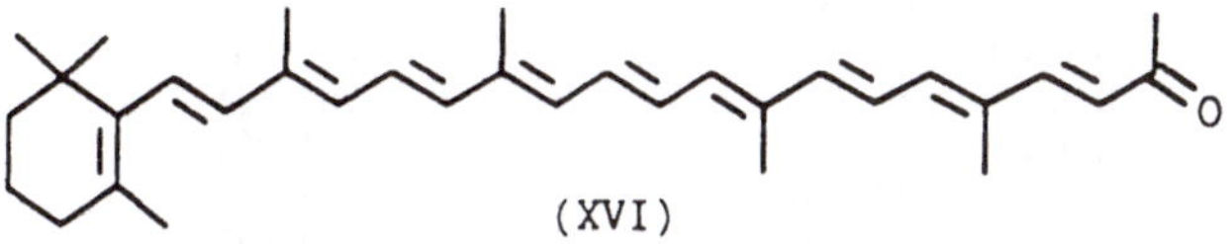

(XVI)

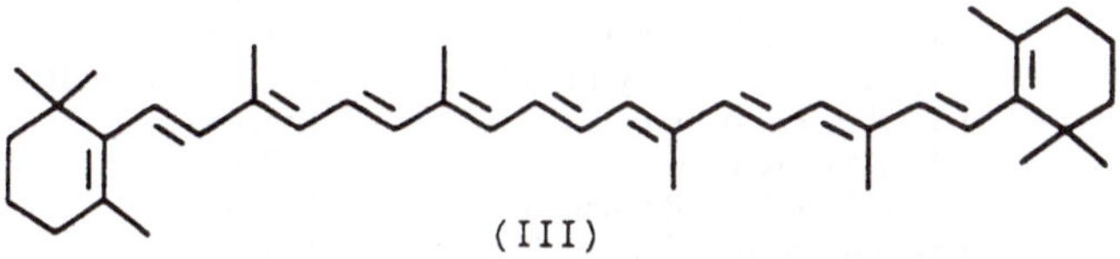

(III)

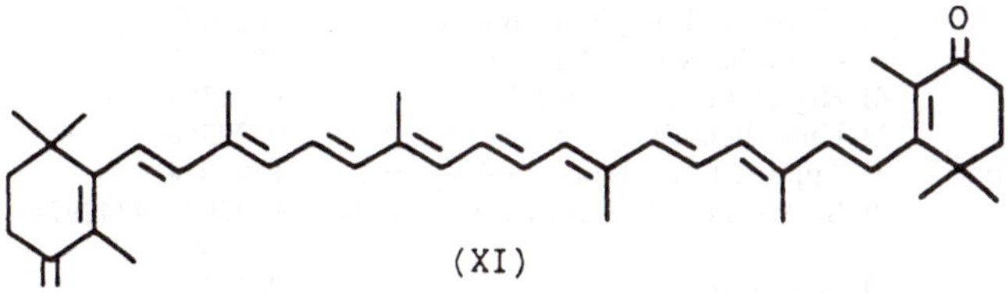

(XI)

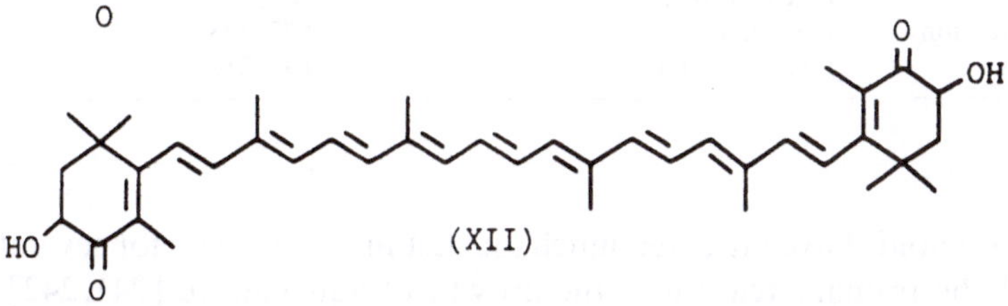

(XII)

Fig. 4. Structures of six synthetic industrially important carotenoids. **XIII:** β-apo-8′-carotenal, (C_{30}); **XV**: β-apo-8′-carotenoic acid ethyl ester, (C_{32}); **XVI:** citranaxanthin, (C_{30}); **III:** β-carotene, (C_{40}); **XI:** canthaxanthin, (C_{40}); **XII:** racemic astaxanthin, (C_{40}). From Ref. [11]

Table 5. Estimated world market for synthetic and biological carotenoids (1992–2000) [235]

	1992 $ million Synth.	Bio.	1996 $ million Synth.	Bio.	2000 $ million Synth.	Bio.
β-carotene						
Food	50	10	75	35	110	60
Nutritional/						
Cosmetic	45	15	95	45	140	130
Astaxanthin/						
canthaxanthin	120	10	210	40	305	150

Table 6. World patents concenring microbial synthesis of carotenoids

Subject	Patenting Company/Individual	World Patent #
1) *P. rhodozyma*	1) Lion Co. (2)	9290687, 9290686
protoplast fusions	2) Phillips Petroleum Co.	9015322
2) Novel *P. rhodozyma*	1) Lion Co.	9554017
strains	2) Gist-Brocades NV	9520648
	3) Enzymatix Ltd.	8637888
	4) Unilever Plc.	8954610
3) *P. rhodozyma*	1) Phillips Petroleum Co.	8815125
fermentation conditions	2) Osaka Chem. Alloy KK	8493807
	3) Biocolours I/S	7673631
4) Astaxanthin feeding	1) Kyowa Hakko Co. Ltd.	9711774
	2) Sanraku Ocean	3650795
	3) Aquatic Diet Techno.	3141032
5) *P. rhodozyma*	1) Villadsen, I.S.	9324671
mutagenesis protocols	2) Pillips Petroleum Co.	8712855
	3) Igene Biotechnology Inc. (2)	8569501, 8196510
6) Astaxanthin	1) Phillips Petroleum Co.	9551253
extraction procedures	2) Ki Kasei KK	9127959
	3) Mikalsen, E.; Mikalsen, G.	8306420
7) Algal astaxanthin	1) Dainippon Ink Chem. KK	8641955
production	2) Commiss. Energie Atomique	7820438
	3) Chlorella Kogyo KK	7993977
	4) Higashimaru Shoyu KK	9443386
	5) Univ. Ben-Gurion Negev Res. & Dev.	9729356
8) Zeaxanthin production	1) Applied Food Biotechnology Inc.	8598054
Novel organisms	2) Kaiyo Biotech. Kenkyusho KK (3)	9742075, 9742974 9418382
	3) Zeagen Inc.	9418382
9) Zeaxanthin production	1) Amoco Co.	8777393
Genetic engineering	2) Kirin Beer KK	8335212

Microbial carotenoids have attracted much interest in recent years for several reasons. One of the primary reasons is the growth of aquaculture [241, 242], currently the fastest growing sector of agriculture. The farming of salmon and shrimp increased greatly during the 1980s. Abount 260000 metric tons of salmon and 700000 metric tons of shrimp were farmed in 1991 in over 40

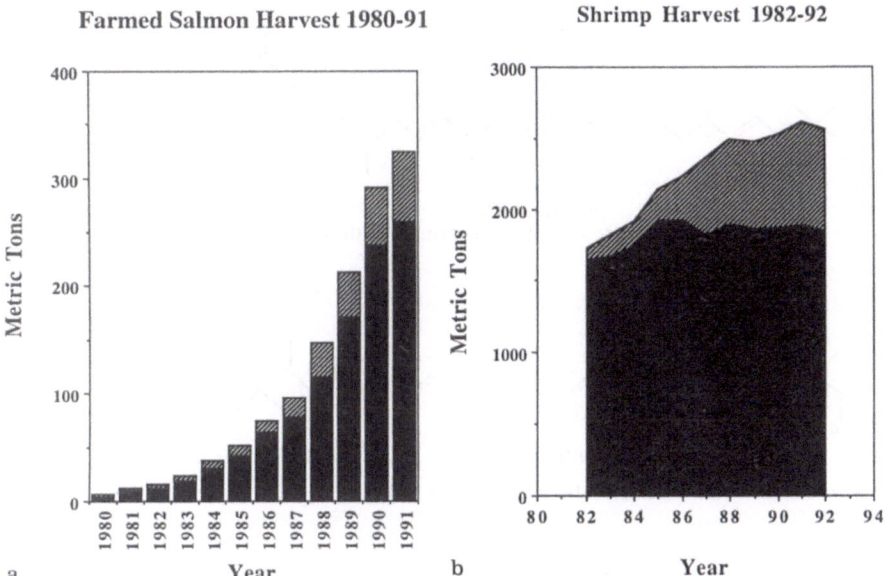

Fig. 5. Worldwide farmed salmon and shrimp crops in thousands of metric tons. **a** (salmon): *solid bars* represent Atlantic salmon, *hatched bars* represent Pacific salmon; **b** (shrimp): *solid area* represents wild-catch, *hatched area* represents farmed shrimp

countries (Fig. 5). Ten years ago, farmed salmon and shrimp supplied 3–5% of the world consumption, but this has expanded to 25–30%. Aquaculture is currently about a $20 billion per year industry which is expected to grow to $40 billion per year by the year 2000. Astaxanthin is an important source of pigmentation in farm-raised salmonids and crustaceans. Carotenoids also are the source of vitamin A in animal diets. Carotenes, particularly β-carotene, have traditionally served this role in animal and human nutrition but other carotenoids including the xanthophylls, zeaxanthin and astaxanthin can be converted to vitamin A in crustaceans and some other animals. Certainly, a primary contributing factor to the increased interest in natural carotenoids is the current trend of avoiding food additives and synthetic chemicals in foods. The structures of carotenoids under investigation for industrial production are shown in Fig. 6.

8.2 Biological Processes for Production of Carotenoids

8.2.1 Lutein

Lutein and lutein extracts are important carotenoid sources of pigmentation of broilers and hens' eggs [239, 243], and sales of pigmenters in the United States was estimated to about $150 million per year [244]. Lutein is also one of the

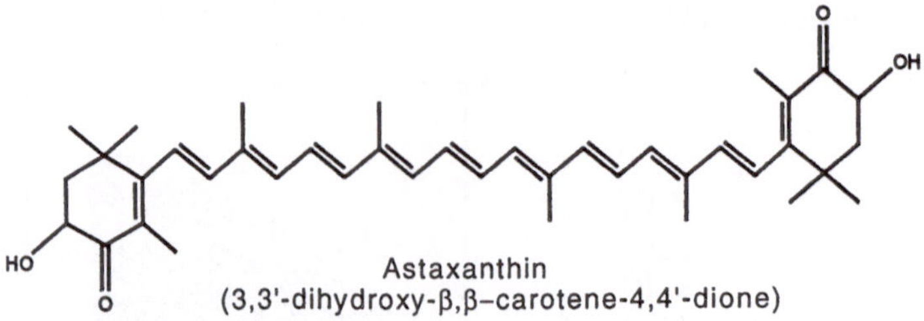

Astaxanthin
(3,3'-dihydroxy-β,β–carotene-4,4'-dione)

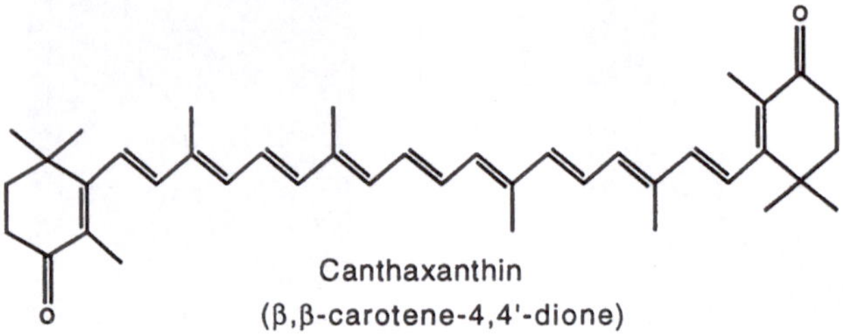

Canthaxanthin
(β,β-carotene-4,4'-dione)

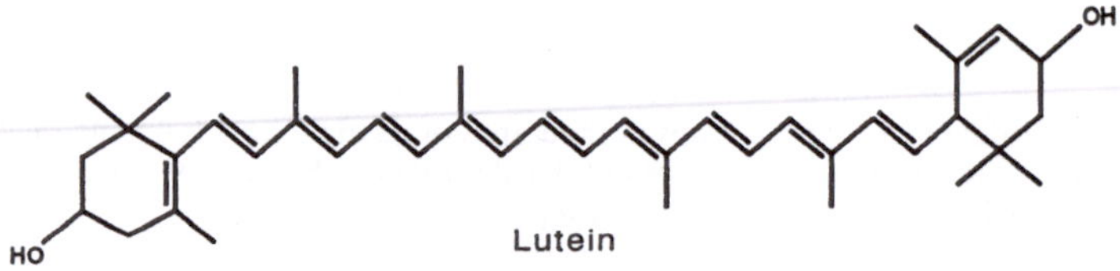

Lutein
(β,ε-carotene-3,3'-diol)

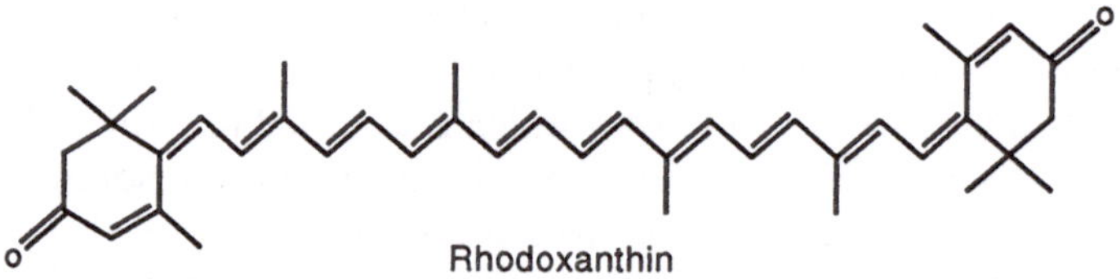

Rhodoxanthin
(4'5'-Didehydro-4,5'-*retro*-β,β-carotene-3,3'-dione)

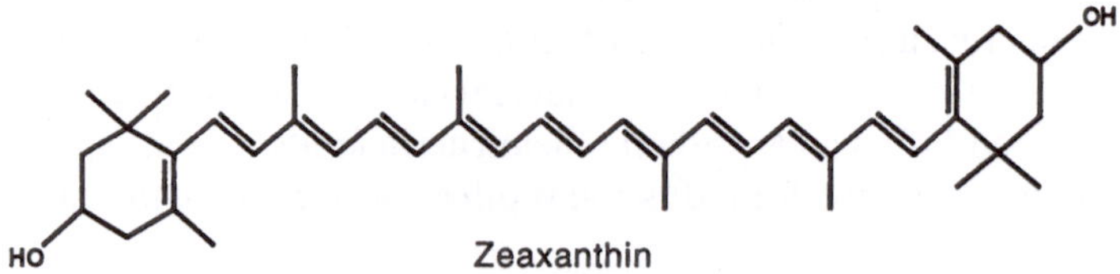

Zeaxanthin
(β,β-carotene-3,3'-diol)

Fig. 6. Structures of some carotenoids under investigation for synthesis by microbial processes

most common carotenoids in freshwater fish, giving pond-reared trout a yellow coloration [245]. For effective egg yolk pigmentation a base pigment of lutein is required. Ketocarotenoids also enhance the color [246]. Generally, a combination of lutein and zeaxanthin is used to give the desired color intensity [246].

With improvements in the efficiency of conversion of feed to body weight the amount of carotenoid ingested in the standard diet containing alfalfa meal or corn meal was often insufficient to provide proper pigmentation [247]. Carotenes are poorly deposited in egg yolks which is reflected in the carotenoid composition of 70% lutein and 30% zeaxanthin [246]. Dehydrated marigold was more effective than meal; the composition of carotenoids in the meal was 64% lutein, 31% antheraxanthin, and 3.5% α-cryptoxanthin. Saponification of the esters to free lutein improved deposition by about 30% [246].

Lutein in alfalfa meal, corn gluten meal, and marigold flowers [248] and concentrates are available commercially for poultry pigmentation. Another commercial source of lutein is marigold petals (*Tagetes erecta*), which are dehydrated, ground and marketed for coloration [249]. Microbial processes for lutein production have been investigated. Patents exist for lutein production in the microalgae *Chlorella*, *Chlorococcum*, *Chlamydomonas* and *Spongiococcum* [250]. *Spongiococcum excentrum* was commercialized for use in chicken feed [250] but it is no longer available. F. Hoffmann-La Roche has patented the production of lutein using *Chlorella pyrenoidosa* cultured at 35 °C, however this was never commercialized [244, 251]. Industrial production of lutein by the blue green alga *Spongiococcum excentrum* by Grain Processing Co. (Muscatine Iowa) is marked as A-Zanth [246]. In this process, *S. excentrum* is grown in a 9 day culture at 28 °C on glucose, corn steep liquor, urea and mineral salts with glucose being added in a feed batch process. This process was reported to yield about 294 mg of lutein per liter of culture.

8.2.2 Zeaxanthin

Zeaxanthin is produced by plants, various algae, cyanobacteria, *Mycobacterium*, *Xanthobacter*, *Erwinia*, and *Flavobacterium*. *Flavobacterium* appear to be among the best sources and several patents for production of zeaxanthin have been filed [252]. *Flavobacterium* cultures generate 10–40 mg of 3R, 3'R zeaxanthin per liter. Supplementation of the media with isoprenoid precursors can increase titers to $335 \, mg \, l^{-1}$ [249]. High liters of zeaxanthin ($15 \, mg \, g^{-1}$) have been obtained in *Flavobacterium* in laboratory batch cultures [253].

Processes for zeaxanthin production have also been developed using the cloned genes for zeaxanthin biosynthesis from *Erwinia* [254]. The clustered genes from *E. herbicola* were initially cloned into *E. coli*, which conferred a yellow color to the organism. Extensive alterations of the genes were necessary for expression in *Saccharomyces cerevisiae*. Gene reconstruction increased the activity of GGPP synthase for 6.35 to $23.4 \, nmol \, min^{-1}$ by deleting portions of the gene. In order for the lycopene cyclase gene to be expressed in *S. cerevisiae*, the original GTG (methionine) start codon had to be changed to ATG. The carotenogenic genes were integrated into the chromosome of *S. cerevisiae*. The gene encoding phytoene synthase was first fused to the phosphglycerate kinase promoter of *S. cerevisiae*, then transformed into the yeast using an integration vector. It was reported that several % zeaxanthin was obtained, and the carotenoids appeared to concentrate in lipid globules in the yeast.

8.2.3 Rhodoxanthin

Rhodoxanthin is distributed widely in leaves, fish and bird feathers [244]. In bacteria it has been detected in a methanol utilizing bacterium, *Pseudomonas extorquens* [255]. The pigment level in *P. extorquens* was found to be growth related, increasing 4-fold during the stationary phase. Nutrient conditions were also found to affect pigmentation. Ethanol and Mg^{2+} inhibited rhodoxanthin biosynthesis. Limitations of Mg^{2+} increased the production 2.3-fold over cultures with excess Mg^{2+} [255]. Use of glycerol and ammonium nitrate as carbon and nitrogen sources gave the highest productivity [255]. Surprisingly, high O_2 supply inhibited pigmentation, and pigment production in an O_2 limited culture was 1.9-fold higher than the yield in excess O_2 [255].

8.2.4 Canthaxanthin

Canthaxanthin was originally isolated from the mushroom *Cantharellus cinnabarinus* and from yam tubers (*Dioscornea*) [246, 256]. Canthaxanthin is currently approved as a food additive in the U.S. and 35 other countries [246]. In addition to other keto-carotenoids such as astaxanthin and echinenone, canthaxanthin is produced in certain green algae (*Haematococcus*, *Chlorella*, *Chalamydomonas*, *Scenedesmus*, and *Ankistrodesmus*), usually in response to stress and nutrient deficiency [257]. Three microbial sources have primarily been studied for commercial canthaxanthin production. *Rhodococcus maris* grown on hydrocarbons had a poor yield [250]. The green alga *Dictyococcus cinnabarinus* under nitrogen deficient conditions produced canthaxanthin, but also with a low yield and slow growth [250]. The most promising source of canthaxanthin appears to be a *Brevibacterium* sp. isolated from oil fields [250]. This process has been optimized for *Brevibacterium* strain KY-4313 by using a high carbon/nitrogen ratio. This involves a two phase process utilizing nitrogen starvation in the second phase, and supplementation of the medium with hydrocarbons and higher alcohols [77]. Since genetic techniques were not available for this organism, strain improvement has relied on random mutagenesis followed by visual screening [250]. Mutagenesis has yielded carotenoid hyperproducing mutants giving 490 $\mu g\,g^{-1}$ cell and 4200 $mg\,l^{-1}$ compared to 370 $\mu g\,g^{-1}$ and 3100 $mg\,l^{-1}$ in the wild type; echinenone levels also increased. A medium considered for industrial application was a hydrocarbon supplemented medium (pH 7.0) containing 2–4% octadecane, malt extract (0.1%), Tween 40 (0.01%) vitamin B_{12} and a high C/N ratio. This formulation yielded canthaxanthin levels of 600 $\mu g\,g^{-1}$ cell, 1–2 $mg\,l^{-1}$ canthaxanthin, and a cell density of 3.5 $g\,l^{-1}$. Mg^{2+} depleted media and substitution of yeast extract for malt extract improved these results, strong aeration was slightly inhibitory. The feeding of carotenoid biosynthesis intermediates or addition of alcohols did not affect pigmentation [77]. In hydrocarbon supplemented media the main limit to growth was the accumulation of toxic carboxylic acids [77]. This led to a production strategy of periodic renewal of the aqueous phase to remove toxic metabolites and periodic octadecane supplementation, which gave a two-fold increase in cell mass as well as increased pigmentation [77]. Because of the difficulties in using hydrocarbon supplemented media, the canthaxanthin synthesis was attempted in a rich medium (Brain Heart Infusion; BHI), which gave increased cell mass but decreased pigmentation [77]. Supplementation of BHI with higher alcohols (propanol, isopropanol or retinol) increased pigmentation. The best results were obtained using 10–20 g of propanol l^{-1}, while retinol as well as vegetable oils gave increased β-carotene levels at the expense of canthaxanthin and decreased cell mass [77]. Dicarboxylic acids (fumaric, maleic and malic acid) stimulated pigmentation. The optimum formulation contained fumaric acid (5%), beet molasses (4%) and sucrose ester (1%). This yielded 9.3 mg of canthaxanthin and 12.6 g of cell mass per 1 [77].

8.2.5 β-carotene

Syntheis of β-carotene in mucoral fungi and green algae has been thoroughly reviewed [258–260]. Industrial methods for β-carotene have been developed using the mucoral fungus *Blakeslea trispora*. Detailed investigations of media composition and culture conditions resulted in a process potentially competitive with chemical synthesis. The US Department of Agriculture process [260] yielded about 17 mg of β-carotene per g mycelium or about 1 g l^{-1}. The process has been improved and yields about 30 mg β-carotene per g mycelium and about 3 g l^{-1} [252]. The medium composition for the improved process using mated cultures of *B. trispora* had certain interesting components which may stimulate production. During the early exponential phase thiamine, $MnSO_4$, soya oil, cotton-seed oil, kerosene, and isoniazid were added. After 48 h, β-ionone, kerosene, and glucose were added and glucose continually until the end of the 185 h process.

Phycomyces also can accumulate very high levels of β-carotene (~ 30 mg g^{-1}) in lipid globules in the mycelium [256]. Production of β-carotene has only been studied in laboratory conditions, usually on solid agar media, but efforts are underway to develop a submerged culture process.

The superior process for microbial production of β-carotene appears to be algal culture using the green, biflagellate green alga, *Dunaliella*. Currently there are at least four commercial operations in Australia, Israel, and the United States. *Dunaliella* powder and extracted β-carotene have been marketed since 1985 [258]. β-Carotene accumulates in oil globules in the interthykaloid spaces of the chloroplast when *Dunaliella* is cultured with intense illumination, a stress temperature, high salinity, and a deficiency in nitrogen or sulfur. In inducing conditions β-carotene reaches levels of up to 14% of the dry weight, while under noninducing conditions the content is about 0.3%. High β-carotene producing mutants have been obtained by selecting for the ability of β-carotene to protect the alga from killing in blue light [261]. Algal β-carotene cannot compete economically with synthesized product, but it has found a niche in "Natural" markets.

8.2.6 Astaxanthin

Astaxanthin is the primary pigment in salmonids and shrimps, which must be supplied in their diet. Currently, most of the astaxanthin is supplied through chemical synthesis. However, astaxanthin is a complex molecule and the synthesis can be a difficult one. Synthetic astaxanthin sells for about $2000 kg^{-1}. Because of the potential market in aquaculture, there has been considerable industrial interest in developing competitive biological sources of astaxanthin.

Currently, biological astaxanthin is not approved as a feed additive in the U.S., forcing the industry to use canthaxanthin. However, it has been shown that canthaxanthin fades faster upon cooking and will fade in fish if feeding is stopped [245]. Consequently, higher canthaxanthin levels are recommended in feeds. For salmonids, $\geq$ 20 µg astaxanthin per g feed is required compared to $\geq$ 40 µg canthaxanthin per g over a two to four month period. It is likely that the Food and Drug Administration will approve astaxanthin in the near future for certain foods and feeds.

Salmon and shrimp differ in their capacity to metabolize carotenoids. Most crustacea deposit carotenoids in the eyes, blood, eggs and carapace [250]. Astaxanthin is the most common carotenoid found, either free, esterified to short chain fatty acids, or noncovalently linked to protein. β-Carotene, lutein, canthaxanthin and echinenone are also often present. Generally crustacea are non-selective and absorb xanthophylls and carotenes equally well. Brine shrimp can convert β-carotene to echinenone and to canthaxanthin but hydroxy carotenoids are not detected [262]. However, when hydroxycarotenoids were provided in the diet to a crustacean (*Daphnia*), they were converted to astaxanthin [250]. In contrast to crustacea, fish are very limited in their ability to metabolically transform carotenoids. Generally, chemical modifications are limited to the esterification of xanthophylls. Esterified compounds usually are deposited in the skin, while the free form is generally found in the flesh, liver, ovary and digestive organs [263].

Astaxanthin has been identified in several microorganisms including the basidiomyceteous fungus *Peniophora* [54], the heterobasidomycetous yeast *Phaffia rhodozyma* [116], the hydrocarbon-utilizing bacteria *Mycobacterium lacticola* and *Brevibacterium* [245], and in various green algae including *Haematococcus* sp, *Neochloris wimmeri* and *Chlamydomonas nivalis* [264]. The higher plant *Adonis aestivalis* species produces astaxanthin in its flower petals, which can be dried and ground without major loss of pigment [265]. Commercialization of *Peniophora* has not been pursued and pigment production by *Mycobacterium* and *Brevibacterium* isolates were poor (~ 30 µg g^{-1} and 3 g l^{-1}, respectively) [245].

The primary biological sources of astaxanthin currently being considered for industrial production are the green alga *Haematococcus pluvialis* and related species, the heterobasidiomyceteous yeast *Phaffia rhodozyma*, and crustacean extracts. With the large quantities of crustacean waste generated each year, much effort has gone into using this as a carotenoid source in fish aquaculture. Generally, the feed requires 20–30% by weight of crustacean waste meal. The major problem in using waste meals is the high level of ash and the low energy content of the shells [266]. Krill oil contains relatively high astaxanthin levels (727–1080 µg g^{-1}), while krill meal contains only 15–200 µg g^{-1} and frozen krill 15–77 µg g^{-1}. In krill, most of the astaxanthin is present as the diester which apparently is not absorbed as efficiently as the free form [266].

Haematococcus pluvialis is a unicellular, motile, green freshwater alga of the class Chlorophyceae. Under favorable growth conditions, *Haematococcus* exists as a single celled biflagellate swimmer capable of photosynthetic autotropic growth [264]. In unfavorable growth conditions cells lose motility and form cysts. This encystment is accompanied by the synthesis and deposition in lipid globules of large quantities of astaxanthin and other carotenoids. Formation begins in the perinuclear region and proceeds outward in the cytoplasm. Astaxanthin can account for up to 5% of the cell mass [15].

During the encystment process, the relative quantities of individual carotenoids change considerably, going from 75–80% lutein and 10–20% β-carotene to over 80% astaxanthin. However, this increase in astaxanthin is not due to conversion of these other carotenoids to astaxanthin [267]. The environmental and nutritional factors which trigger encystment and carotenogenesis have been investigated in *H. pluvialis*. Nitrogen starvation occurs in the same period as encystment although the two processes were reported not to be physiologically linked [268]. Encystment is also triggered by the addition of NaCl (up to 1%) [268]. Astaxanthin accumulates when cell division is inhibited, and phosphate but not nitrogen starvation and high light intensities also lead to accumulation of $\leq 3\%$ astaxanthin [269]. A study of nutritional requirements has shown that KNO$_3$ (0.5 g l^{-1}) is the best nitrogen source for pigmentation, and intermediate levels of phosphate (0.001–0.2 g l^{-1}) are optimum. Encystment is triggered by exhaustion of nitrogen and increasing phosphate levels [268]. Iron (Fe^{2+}) stimulates carotenogenesis, but only in the presence of acetate [270]. It was suggested that Fe^{2+} promotes the Fenton reaction which generates hydroxyl radicals. Exposing the cells to active oxygen species O$_2$-, OH·, ^{1}O$_2$ and H$_2$O$_2$ in the absence of Fe^{2+} induces the production of astaxanthin [270], which is a potent quencher of active oxygen species. Carotenoid levels are increased by the feeding of Fe^{2+} and acetate following the cessation of growth in a two stage process [271]. Furthermore, it was found that activation by Fe^{2+} or H$_2$O$_2$ is proportional to light intensity. In the absence of light, no activation by Fe^{2+} or

H_2O_2 was observed. Light activation occurred more than 18 hours after acetate addition [272]. Technologies have also been investigated to maximize cell growth and density prior to encystment, as well as achieving the maximum levels of astaxanthin after encystment. Lwoff and Lwoff [273] first reported in 1930 that addition of acetate to mineral peptone media enhanced carotenogenesis. Goodwin and Jamikorn [267], however, suggest that acetate is responsible for increased growth but not increased pigment. More recently, Borowitzka et al. [268] demonstrated that acetate increases pigmentation in addition to growth. Although light is not strictly required for pigmentation, illumination gives 7-fold higher pigmentation and two-fold higher growth [274]. Thiamine was also found to stimulate growth.

The astaxanthin biosynthetic pathway has been investigated in *H. pluvialis*. During encystment, astaxanthin synthesis is not synthesized from pre-existing lutein and β-carotene pools [267]. Radiolabeling studies showed that exogeneous sources of carbon are a major source of ketocarotenoids and that there is no evidence for the conversion of β-carotene to astaxanthin [275], although echinenone and canthaxanthin were proposed as astaxanthin precursors [275]. This was supported by the increased demand for carbon in cells during encystment than during vegetative growth. Feeding cells during encystment with carotenoid precursors much as mevalonic acid, pyruvate, malonate or dimethylacrylate increased pigmentation [275].

Microbio Resources Inc. of San Diego, California, has developed a commercial process for the production of astaxanthin from *Haematococcus pluvialis*, marketed as Algaxan Red. *H. pluvialis* is grown in modified Bolds basal medium (pH 7.3). Scale up of the cultures involves a series of 10-fold increments, each stage requiring 5–7 d. The final production stage is in 4500 m³ ponds with encystment being induced by nitrogen starvation or NaCl addition when a cell density of $3-6 \times 10^5$ cells per ml is obtained. Because production occurs in open ponds, contamination is a major obstacle. Contamination generally occurs by wild algae or protozoan predators. In the latter case, predation can eliminate up to 90% of *H. pluvialis* within 72 h.

Following harvesting of the cells, astaxanthin must be liberated from the cysts and protected from oxidation. This is accomplished by grinding the dried cysts to a diameter of $<10 \mu m$ at $-170\,°C$ in the presence of antioxidants such as butylated hydroxytoluene or ethoxyquin [264]. The powder, actually a mixture of 1% free astaxanthin (in the 3S, 3'S configuration), 76% astaxanthin monoester, 7% astaxanthin diester, 1% β-carotene, 7% lutein and 2% violaxanthin [276], is recommended for use at 25 to 100 μg astaxanthin per g salmon feed. Although costs vary, it is sold at $20 kg^{-1} for a 1% astaxanthin product. This is competitive with pure synthetic astaxanthin sold at $2000 kg^{-1} [264].

Another microbial source of astaxanthin is the yeast *Phaffia rhodozyma*. The natural habitat of the orange-red heterobasidiomyceteous yeast *P. rhodozyma* is slime fluxes of deciduous trees including birch, beech, alder, aspen, and dogwood in mountainous regions of Japan, the Pacific Northwest, and Russia [114]. Soon after its initial isolation in 1967, the yeast was recognized as representing a new genus "Rhodozyma". Later Golubev et al. [277] isolated 67 strains from birch fluxes near Moscow. The yeast was formally described as *Phaffia rhodozyma* in 1976 in recognition of the lifelong contributions of Herman J. Phaff to the biology of yeasts [115]. Characteristic properties of *Phaffia* including its repetitive budding from a single cellular site, presence of coenzyme Q-10, xylose in the cell wall, a positive diazonium blue B staining reaction, characteristic 18s rRNA sequences and production of carotenoid pigments. These characters indicate its affinity in the heterobasidiomycetes in the Cryptococcaceae [278–280]. Further evidence linking *Phaffia* to the Crypto-coccacacae is the presence of killer activity. Double-stranded RNA (dsRNA) particles of molecular sizes ranging from 4.3 to 0.75 kilobase pairs were isolated. Strains containing these dsRNA were found to kill strains without dsRNA [281]. Gueho et al. [279] suggested that based on its 26S and 18S rRNA partial sequences *Phaffia* evolved from a fungus having a primitive dolipore such as in *Cystofilobasidium capitatum*. In exception to this proposal, Prillinger et al. [282] suggested that *Phaffia* was related to the Tremellaceae (*Tremella, Trimorphomyces*) which are Hymenomycetes that have complex septa. This is interesting in light of the fact that Tremellales form fruit-bodies that are brightly colored, and the basidiospores germinate by yeast-like budding and in culture the fungus grows in a yeast-like manner [283]. The Tremellaceae grow on stumps of various tress. *Tremella foliaceae* was shown [284] to be closely related to *Filobasidiella neoformans*, which is the perfect stage of the yeast *Cryptococcus neoformans*. The Dacrymycetales, closely related to the Trenellacea, include the jelly fungi whose cells can become packed with oil globules containing carotenes [283]. This taxonomic information suggests that *Phaffia* may have evolved from the Tremellaceae in the higher basidiomycetes and that it may be useful to investigate Tremellales as a new source of astaxanthin.

The carotenoid intermediates found in *P. rhodozyma* are quite different from those found in *H. pluvialis*. *H. pluvialis* contains β-carotene, lutein, violaxanthin, and astaxanthin, which has the 3*S*, 3′*S* stereoconfiguration [273]. In contrast, *P. rhodozyma* produces the 3*R*, 3′*R* configuration of astaxanthin [117] and contains several ketocarotenoids including echinenone, phoenicoxanthin, can-thaxanthin, and HDCO [116]. The difference in configuration is intriguing and suggests distinct enzyme mechanisms or differences in the biosynthetic path-ways. Bjerking suggested [285] that addition of oxygen at the C-4 position before oxygen addition at the C-3 position results in formation of the 3*R* astaxanthin configurational isomer while oxygen addition at the C-3 position before C-4 oxygen addition gives the 3*S* isomer.

The wild isolates of *P. rhodozyma* contain relatively low quantities of caro-tenoids ($\leq$ 500 μg total carotenoids per g yeast), of which 40 to 95% of the overall mixture is astaxanthin [116]. To develop an industrial process for astaxanthin production, it was necessary to develop strains of *P. rhodozyma* with a minimum astaxanthin content of 3000 μg g^{-1}. A number of genetic methods have been utilized to attain this goal. Mutagens have been utilized in *P. rhodozyma* strain development, with ethylmethyl sulfonate (EMS) and *N*-methyl-*N*′-nitro-*N*-nitrosoguanidine (NTG) being the most effective [286]. The first few rounds of random mutagenesis and visual screening yields mutants with significantly elevated pigment levels and carotenoid concentrations of $1500 \, \mu g \, g^{-1}$ are relatively easily obtained [286]. However later rounds of mutagenesis give smaller pigment incremental increases (0–10%). Beyond pigment levels of $\sim 1200 \, \mu g \, g^{-1}$ it is very difficult to detect higher pigment levels by visual inspection possibly due to oxygen transfer limitations on plates [231, 287]. Strain instability and the tendency of mutants to produce a lesser proportion of astaxanthin in the total complement of carotenoids is also a significant problem. This has led our laboratory and others to investigate positive genetic selection systems for carotenoid hyperproducing strains. Carotenoid synthesis in several carotenogenic fungi is subject to end product inhibition (e.g. β-carotene in *Phycomyces*) and it seemed feasible that analogs to carotenoids could be used to select analog resistant mutants with altered pathway regulation. Chun et al. [288] suggested that β-carotene feedback inhibits carotenoid synthesis in *P. rhodozyma* in a manner similar to *Phycomyces*. Lewis et al. [287] used the β-carotene analog β-ionone as an inhibitor and was able to isolate hyperproducing strains in *P. rhodozyma*. β-Ionone (10^{-4} M) decreased pigmentation 5-fold and killed $\sim 20\%$ of wild-type cells whereas the hyperproducing strains were resistant to 10^{-3} M β-ionone. This difference in viability was exploited as a selection system in conjunction with NTG mutagenesis to produce mutants with $1000 \, \mu g \, g^{-1}$ pigmentation after 2 rounds. However, while β-ionone was effective under these conditions, it was concluded that β-ionone was effective only in initial mutation cycles [289]. Schroeder and Johnson [290] have demonstrated that the degradation of astaxanthin in live cells by tert-butylhydroperoxide resulted in an ~ 5-fold increase in β-carotene levels. The accumulation of β-carotene with decreased intracellular astaxanthin suggests that astaxanthin may be involved in feedback regulation. Astaxanthin analogs may be more effective than β-carotene analogs in selections for hyperproducing strains.

A second strategy for positive selections was to exploit the antioxidant activity of carotenoids. Chang [291] initially demonstrated hyperproducing cell lines to be more sensitive to H_2O_2. However, Schroeder and Johnson [292] showed an age dependence in which young hyperproducing cultures were more resistant to H_2O_2 than wild-type cultures. Use of this as a selection gave slightly increased pigmentation after repeated challenges [292]. A second oxidant which is specifically detoxified by carotenoids is $^1O_2$ [293]. Singlet oxygen is an excited electronic state of O_2 which when it decays to its ground state causes the peroxidation of lipids, cellular membrane damage, and damage to DNA. Singlet oxygen is relatively long-lived, especially in lipid environments, and its diffusion radius in cells has been estimated as $100 \, \text{Å}$ [294]. The lipid peroxidative activity of $^1O_2$ is particularly effective because of the increased lifetime of $^1O_2$ in membranes compared to the aqueous phase ($25–100 \, \mu s$ vs $4 \, \mu s$) [295]. Fatty acid analysis of *P. rhodozyma* showed that it is rich in fatty acids which are particularly sensitive to the damaging effect of $^1O_2$. Linoleic and oleic acids, which are quite sensitive, are the two most prevalent fatty acids in *P. rhodozyma* (Table 7). Carotenoids are well recognized as being able to detoxify $^1O_2$ catalytically,

Table 7. Comparison of the composition of the *Phaffia* products from Red Star (Universal Foods, Milwaukee, Wisconsin, USA and Natupink Gist-Brocades. Delft, The Netherlands)

Values reported as Red Star/Natupink

Bulk composition

Astaxanthin (ppm)	Protein (%)	Lipid (%)	Ash (%)	Mositure (%
3000/2500	22/32	23/28	3/5	5/5

Amino Acids (%)	Fatty Acids (%)	Vitamins (ppm)	Minerals (ppm)	
Glu: 7.9/9.7	Linoleic (18:2n6)	Niacin	Calcium	
Asp: 6.3/6.7	39.65/36.5	1520/ND	248/1900	
Arg: 6.3/6.7	Oleic (18:1n9)	Riboflavin	Sodium	
Ala: 5.6/4.8	32.43/33.2	100/ND	2770/4000	
Leu: 5.1/5.0	Palmitic (16:0)	Vitamin D$_2^*$	Zinc	
Lys: 4.7/4.8	13.18/12.7	85/ND	14/50	
Thr: 3.9/3.9	Stearic (18:0)	Biotin	Phosphate	
Ser: 3.7/4.6	5.62/9.8	1/ND	16200/15000	
Val: 3.7/4.1	Linolenic (18:3n3)	Thiamine	Copper	
Pro: 3.6/3.4	1.26/1.5	14/ND	ND/15	
Gly: 3.6/3.9	Heptadecanoic (17:0)	Pyridoxine	Manganese	
Ile: 2.9/4.2	1.24/1.2	12/ND	ND/11	
Phe: 2.8/2.2	Behenic (22:0)	Pantothenate	Magnesium	
Tyr: 1.9/2.2	0.76/1.1	34/ND	1489/700	
His: 1.7/1.6	Eicosanoic (20:0)	Folate	Potassium	
Trp: 0.7/ND	0.75/ND	6/ND	4135/10500	
	Palmitoleic (16:ln7)		Iron	
	0.4/ND		31/200	

*IU/100g
ND: not determined

and a single β-carotene molecule can detoxify 250–1000 molecules of $^1O_2$ [295]. Carotenoids differ in their capacity to detoxify $^1O_2$, and lycopene and astaxanthin are more effective than β-carotene [296, 297]. Consequently, mutants of *P. rhodozyma* with increased astaxanthin would be expected to be more resistant to damage by $^1O_2$. When mixed cultures of wild-type and hyperproducing strains were cultured together and exposed to light activated $^1O_2$ generators, a monoculture of the hyperproducing line was selected [298]. Singlet oxygen can also be generated in the cell by chemical reactions such as the spontaneous dismutation of O_2^- or by the reaction of H_2O_2 with O_2^- (Haber-Weiss reaction) [299]. This suggests that protection of cells by carotenoids agianst H_2O_2 or O_2^- may be through a $^1O_2$ intermediate. These results indicate that resistance to $^1O_2$ may provide a basis for a positive genetic selection for astaxanthin and also that the function of carotenoids in *P. rhodozyma* is to protect against $^1O_2$ in its natural environment.

Because *P. rhodozyma* is imperfect and pedogamic sexuality has only been recently demonstrated, it has not been possible to perform conventional crosses and genetic analyses. Recombinant DNA technology has been attempted with *Phaffia* but not published. One of the most powerful methods of obtaining recombinant strains in basidiomycetes is the development of transformation protocols [300]. Genetic transformations have been developed for several basidiomycetes including the yeasts *Cryptococcus neoformans* [301] and *Rhodosporidium toruloides* [[302]. van Ooyen [303] has reported a transformation system for *P. rhodozyma*. An actin sequence which hybridized with *Phaffia* genomic DNA was used to isolate a DNA fragment containing a *Phaffia* promoter. A marker gene encoding phleomycin resistance was fused with this promoter and used to transform *P. rhodozyma*. The development of this system should enable the expression of heterologous proteins in *Phaffia*, including enzymes catalyzing carotenoid biosynthesis.

Protoplast fusion has also been used for strain development of *P. rhodozyma* [288]. Previously it had been reported that the generation of auxotrophic mutants was difficult, possibly indicating polyploidy or aneuploidy in *P. rhodozyma* [304]. Chun et al. [288] were able to isolate strains with auxotrophic markers for tryptophan, leucine, methionine and arginine and suggested that *P. rhodozyma* was haploid. Using these auxotrophs, 30 hybrids were generated, three of which had significantly higher astaxanthin levels than the parents (2000 compared to 1600 μg g^{-1}). The hybrids were found to have 2-3-fold more DNA than the parents, were uninucleate indicating nuclear fusion, and were stable (segregation frequency $< 2.1 \times 10^{-4}$ reversion). Prevatt [305] also reported generating *P. rhodozyma* fusion strains which produced 4500–8600 mg of astaxanthin 1^{-1} of culture broth. They fused early stationary phase spheroplasts for 15 min using polyethylene glycol as a stabilizer.

Adrio et al. [306] found the isolation of auxotrophic mutants more difficult. They were unable to isolate any auxotrophs from 90 000 EMS- generated or 50 000 NTG-generated mutants. Enrichment for auxotrophs by nystatin yielded mutants at a frequency of 1×10^{-3} after NTG mutagenesis. Nystatin gave 150-350-fold selection. The largest class of auxotrophs required adenine. It was possible to isolate *Ura3* mutants on 5-fluoroorotic acid (5-FOA) plates and then to complement the mutation using a plasmid containing the *Ura3* gene [306]. It has also been the experience of our laboratory that auxotrophs are difficult to isolate in *P. rhodozyma*. This is puzzling since color mutants (white, yellow, red) occur frequently. Aneuploid strains have been isolated in *Phaffia* which express color phenotypes [307]. Aneuploidy in a polyploid strain could explain the difficulty in obtaining some classes of mutants.

As new strains of *P. rhodozyma* are developed it will become increasingly important to demonstrate the genetic uniqueness of the strains as well as confirming that strains used in production do not spontaneously revert during mass culture. Fingerprinting of DNA by restriction fragment length polymorphisms (RFLPs) is a common laboratory technique, and a modification of this strategy was used for typing *P. rhodozyma* [308]. The method used randomly amplified polymorphic DNAs (RAPD) generated with a single oligonucleotide primer containing an arbitrary sequence. Using the PCR template, random regions were amplified from the genomic DNA. Between 3 and 5 major DNA fragments were amplified, which varied in size from 0.7 to 2 kilobase pairs. The banding pattern on using contour-clamped homogeneous electric field (CHEF) electrophoresis differed according to the strains. The analysis suggested that extensive mutations had occurred in the hyperproducing strains. This method could be of use for determining the gentic relatedness and uniqueness of strains.

Varga et al. [309] studied relatedness among 6 wild-type *Phaffia* strains. Relatedness was quantified using a combination of RFLP, RAPD, isoenzyme patterns and electrophoretic karyotyping. It was determined that although *Phaffia* was indeed distinct from other Crytococcaceae, there was considerable heterogeneity among strains. While isoenzyme profiles were the most stably maintained, RAPD analysis was the most sensitive to interstrain differences, even more sensitive than electrophoretic karyotyping.

Fluorescence Activated Cell Sorting (FACS) has been used as a strategy to isolate mutants that overproduce autofluorescent cellular components. This strategy was demonstrated to be successful in isolating algal cells based on chlorophylll autofluorescence [310] and anthocyanin producing mutants of *Aralia cordata* [311]. An et al. [312] investigated this novel screening method for carotenoid hyperproducers based on the autofluorescence of carotenoids. When excited by laser light of 488 nm, carotenoids fluoresce at a characteristic wavelength. Conditions were developed in which autofluorescence was due to cellular carotenoids. Using FACS, it was possible to individually evaluate and sort up to 4500 cells per second for increased fluorescence and consequently increased carotenoid content. When a mixed culture of a wild-type and carotenoid hyperproducer was sorted out by FACS, a 9- to 147-fold enrichment of hyperproducer was attained. NTG mutagenesis followed by cell sorting yielded a 10 000-fold improved efficiency for isolating hyperproducing mutants in a population comprised mostly of wild-type cells [312]. This method could be used to isolate highly fluorescent cells if the optimum sorting windows can be devised and the viability of hyperproducing strains can be maintained.

Confocal fluorescence microscopy and transmission electron microscopy were used to study the subcellular distribution of carotenoids in *P. rhodozyma* [313]. Carotenoids were localized in the external membrane regions and in lipid globules but were absent from the mitochondria. Using fluorescence as an indication of carotenoid content in individual cells, an estimate was made of the maximum cellular content of carotenoids in *P. rhodozyma*. Log fluorescence as a function of carotenoid content gave a linear relationship, and the most highly fluorescent cells contained approximately 15 000 μg astaxanthin g^{-1}, significantly above a commercialization target of 3000 to 5000 μg g^{-1} [59]. However, since several carotenoids fluoresce, particularly phytoene and phytofluene, the content of astaxanthin is an unknown proportion of the total.

A linear reaction network mathematical model of astaxanthin production has been developed [314]. The model supported the hypothesis that the current yield of astaxanthin by high producing strains (3000 to 5000 μg g^{-1}) is much less than is theoretically possible. If no constraint is applied to the formation or consumption of ATP, and certain other assumptions are made regarding the chemical species in the astaxanthin pathway, then the maximum production of astaxanthin from glucose is (in moles) :

$$12 \text{ glucose} + 4 \text{ O}_2 + 14 \text{ NADPH} + 48 \text{ NAD}^+ = 1 \text{ astaxanthin}$$
$$+ 32 \text{ CO}_2 + 14 \text{ NADP}^+ + 48 \text{ NADH} + 12 \text{ H}_2\text{O} + 34 \text{ H}^+ \tag{1}$$

The overall reaction is the linear combination of the constituent reactions with assumed steady state rates. The rates are considered relative, for instance, if the reaction catalyzed by the C_{40} dehydrogenase proceeds at a specific rate, then the reaction catalyzed by HMG-CoA reductase must proceed at ≥ 8 times that specific rate. It is clear that a large quantity of reduced NADH is produced in the reaction which must be balanced. If the cells used part of the glucose as an electron sink, and the coefficient of oxygen is not greater than 4, then the net reaction maximizing astaxanthin production would be:

$$3.83 \text{ mol glucose} + 0.954 \text{ mol ammonia} + 4 \text{ mol O}_2 = 0.213 \text{ mol}$$
$$\text{astaxanthin} + 0.954 \text{ mol biomass } (C_8H_{18}O_4N) + 6.82 \text{ mol CO}_2$$
$$+ 12.67 \text{ ml H}_2\text{O}. \tag{2}$$

This pathway model suggests that the ratio of astaxanthin to biomass could be 2 to 9, tremendously higher than the ratio of 1 to 3300 obtained in wild-type cells. Furthermore, this model shows that 21 moles of O_2 are required per mol astaxanthin produced, and that 1.75 mol O_2 are required per mol glucose consumed for astaxanthin production to maintain redox balance compared to 1.53 O_2 consumed per mol glucose for aerobic biomass synthesis. This model infers that O_2 availability is a major bottleneck in astaxanthin fermentations, and that as O_2 becomes limiting more glucose is diverted to biomass at the expense of astaxanthin formation. It may be possible to reduce the oxygen limitation by providing the cells with an alternate electron acceptor to oxygen in order to oxidize reduced species generated in the astaxanthin pathway. If an alternative oxidant was used, then the cell would require only 1/3 molecule of oxygen per glucose for astaxanthin synthesis, greatly reducing the oxygen transfer problem. The model also infers that hyperproducing mutants with high growth rates may not be easily attained since biomass production is favored by conditions that discourage carotenogenesis [314]. This conclusion is born out by experimental evidence; many, if not all of the hyperproducing strains of P. rhodozyma grow more slowly than their parents.

The modeling analysis suggests that nutrient control and feeding during fermentation are critical for optimal synthesis of astaxanthin and biomass. Initial studies by Johnson and Lewis [315] showed than in wild-type strains as batch cultures entered stationary phase (30–40 h) xanthophyll levels increased over 2-fold and continued to increase up to 128 h. Meyer and DuPreez [316] reported that in batch culture pigment formation slowed after exhaustion of the carbon source. During exponential phase total pigment was produced at a rate of 28.8 $\mu g\,g^{-1}\,h^{-1}$ which slowed to 11.1 $\mu g\,g^{-1}\,h^{-1}$ during stationary phase.

Pigment levels on a dry weight basis increased because of the decrease in dry cell mass that occurs in *Phaffia* cultures during stationary phase. The pH of the medium only slightly affected final cell mass over the range 3.8–7.5. The growth rate was fastest at pH 5.8 and maximum astaxanthin was produced at pH 5.0. pH 4.5 was suggested to be optimal in fermentors. Buffers in this pH region were studied, lactate being unsatisfactory because it was metabolized. The best buffer was potassium hydrogen phthalate. At lower pHs (3.5) the carotene intermediate β-zeacarotene accumulated as shown earlier by Johnson and Lewis [315] indicating that cyclization in the carotenogenic pathway could be altered by stressful conditions. A primary limitation in the yeast astaxanthin culture is low temperature range for growth. Temperature affected growth and astaxanthin formation and the highest specific growth rate ($0.12\,h^{-1}$) was at 22 °C. Bleaching of the cells occurred at higher temperatures. A slightly lower temperature of 15–20 °C was recommended for astaxanthin production [317]. Polulyakh et al. [318] reported that growth of *P. rhodozyma* at 30 °C induced the production of torulene and torularhodin. Other laboratories have not reported that *Phaffia* can grow above 27 °C, and it is possible that their culture was contaminated with *Rhodotorula*. *Rhodotorula* has been found to produce predominantly carotenes at lower temperatures and synthesize torulene and torularhodin as the temperature is raised [316].

The carbon source and its feeding during growth has a pronounced effect on astaxanthin production by *P. rhodozyma*. Glucose at high concentrations inhibited carotenogenesis as well as the maximum specific growth rate (μ_{max}) [316]. Consequently, fed batch cultures are usually carried out, which avoids repression of carotenogenesis early in the process. Disaccharides, such as sucrose, maltose and particularly cellobiose were best for pigmentation while sucrose and glucose gave the best growth. Good pigmentation on cellobiose, which is probably slowly hydrolyzed compared to sucrose, suggests that astaxanthin biosynthesis may be regulated by catabolite repression. High concentrations of glucose ($50\,g\,l^{-1}$) increased carotenes and decreased astaxanthin [315]. Prevatt [319] suggested that astaxanthin levels and foaming increased with carbon limitation. Conversely, high levels of sugar in the medium, particularly in lag or early exponential phase, caused decreased astaxanthin levels and the production of alcohols and aldehydes. A 12-fold increase in β-carotene levels was found at high sugar (5%) levels [315]. Similarly, it was reported [320] that β-carotene and β-zeacarotene were the main pigments in cells cultured in media containing $100\,g\,glucose\,l^{-1}$. The high levels of carotenes and formation of ethanol and glycerol in media containing high sugar was proposed to be caused by limitations in O_2 uptake. A careful study of sucrose utilization and astaxanthin biosynthesis was carried out in batch culture [316]. It was reported that the uptake of the sucrose monomers was rate-limiting for growth. Sucrose was hydrolysed by invertase to its constituent monomers within 8 h. However, the resulting glucose was not completely utilized until 29 h and fructose uptake was delayed by the presence of glucose. Ethanol ($17\,g\,l^{-1}$) and glycerol ($2\,g\,l^{-1}$) were detected as metabolic products at high sugar concentrations. These results emphasize that it is critical to control availability of sugar during *Phaffia* processes.

Nitrogen sources also affect growth and pigmentation in *P. rhodozyma*. Peptone was reported to be the best single source of nitrogen for pigmentation [317]. However, a mixture (1: 1: 1) of yeast extract, beef extract, and KNO_3 at a nitrogen concentration of $5 g l^{-1}$ protein gave the best pigmentation. The positive effect of KNO_3 is puzzling since *Phaffia* cannot use nitrate as an N source. The use of yeast extract as a nitrogen source increased carotenoid levels [315, 316]. In another study, two-fold higher astaxanthin content was obtained with yeast extract [62]. Culturing cells under N limitation decreased the astaxanthin content of *P. rhodozyma* [320]. This is surprising since a high C/N ratio generally increases lipid biosynthesis in fungi. However, it is possible that the common substrate acetyl-CoA is diverted from carotenoid biosynthesis to lipid formation. The highest astaxanthin levels in batch cultures were observed when no residual ammonia remained in batch cultures. An et al. [286] observed that nitrogen was assimilated more slowly in hyperproducing mutants of *P. rhodozyma*, possibly indicating that astaxanthin biosynthesis might be subject to nitrogen regulation.

The synthesis of astaxanthin by *P. rhodozyma* was investigated in continuous culture [316]. The optimum conditions for pigment synthesis was carbon limitation. The total carotenoid and astaxanthin levels decreased with increasing dilution rate, again supporting that carbon limitation is very important for pigment synthesis. Even at the lowest dilution rate investigated ($0.043 h^{-1}$), the carotenoid content was about half that obtained in batch cultures with the strain. An increase in the glucose concentration in the chemostat decreased μ_{max}, $Y_{x/s}$, total pigmentation and astaxanthin. An attempt to determine the maintenance energy coefficient was not possible since carotenogenesis is energy consuming, and varies with the dilution rate. The investigators determined the yield ($Y_{x/s}$) of *P. rhodozyma* (mutant J4-3) on various substrates (g cells per g substrate): N, 17.7–19.62; PO_4^{2-}, 87.2; K, 65.8; Mg, 1040; Ca 2687 [316]. The yield on N was at the upper range for yeasts, suggesting that nitrogen utilization is related to pigment accumulation. It would be of value to determine the yield on N for wild type yeasts as well as hyperproducers. Continuous culture appears to be good approach to determine the effect of limiting nutrients (e.g. P, N, Mg) on astaxanthin synthesis but it is limited by the low dilution rates and poor pigment synthesis. Because astaxanthin production is highest at low dilution rates and is repressed under conditions that promote growth, continuous cultivation would be sensible only on an industrial scale, with an integrated process layout [316].

Oxygen is a critical nutrient affecting astaxanthin synthesis. At aeration rates of $<30 mmol O_2 l^{-1} h^{-1}$ β-carotene was the primary carotenoid in *P. rhodozyma* [319]. The combination of 4% glucose and $5 mmol O_2 l^{-1} h^{-1}$ aeration gave pigment levels of only $30 \mu g g^{-1}$ [315]. Our laboratory has noted that *P. rhodozyma* switches from cyanide-sensitive to cyanide-insensitive respiration on entry into stationary phase and that carotenogenesis also increases after this transition. Cyanide-insensitive respiration has been associated with generation of O_2^- [321]. These data support that oxygen species such as O_2- or $^1O_2$ are involved in carotenoid regulation in *P. rhodozyma*.

Although carotenoid synthesis is many fungi is stimulated by light, illumination was initially reported to have no effect on pigmentation [315]. Later studies by An and Johnson [322] using higher light intensities showed that light depressed both growth and pigment formation with blue light having the strongest effect. Light also increased β-zeacarotene levels, indicating cellular stress. Carotenoid hyperproducing mutants were less apt to become bleached by illumination. The combination of the respiratory inhibitor antimycin and light strongly increased pigment levels in *P. rhodozyma*. Light also restored growth in cells inhibited by antimycin after a 12–24 h lag which suggested that light might induce or feed an alternate respiratory system that facilities NADPH oxidation and continued growth. Salicylhydroxamic acid (SHAM) and propyl gallate (PG), which are inhitiors of antimycin-insensitive respiration in fungi, blocked induction of growth in antimycin by light. These data supported that an antimycin-insensitive oxidase is present in *P. rhodozyma* which affects carotenoid production. It is intriguing that partially purified alternative oxidase (ubiquinol oxidase) preparations from plants contain the presence of carotenoids and the oxidase [323] suggesting they may be physically associated with carotenoids in the cell.

Exposure of cells to antimycin for extended periods gave rise to hyperproducing colonies which were more sensitive to antimycin, and had altered respiratory function [286]. Surprisingly, antimycin-sensitive mutants that produced higher levels of pigment showed increased sensitivity to photokilling, particularly by blue light. It is likely that photoreactive molecules other than carotenoids such as flavins or porphyrins are more associated with light sensitization in *P. rhodozyma*, as has been postulated for other microorganisms [324]. The data also imply that the endogenous sensitizers, flavins or porphyrins, are involved in regulating carotenoid biosynthesis. A yellow mutant, yan-1, which produces β-carotene under normal culture conditions, formed xanthophylls when treated with antimycin + light. These data suggest that induction of an alternative oxidase system, possibly a flavin oxygenase or cytochrome P-450, allows growth in antimycin-inhibited yeasts and also stimulates carotenoid biosynthesis. Preliminary work in our laboratory has shown that in the presence of the potent cytochrome P-450 inhibitors, metyrapone and piperonyl butoxide, *P. rhodozyma* grows well but is nonpigmented. These data indicate a role for cytochrome P-450 in carotenogenesis. Multiple plant cytochrome P-450s are well recognized in plants which are involved in hydroxylations of isoprenoids [325].

In industrial scale productions, costs could be reduced by the use of non-refined carbon sources. Beet and cane molasses have been investigated for production of *P. rhodozyma*. Haard [326] grew *P. rhodozyma* on 7–10% molasses and showed 2–3 times higher astaxanthin higher astaxanthin production than on simple sugars. With 10% molasses, astaxanthin production was 3 times that of a corresponding amount of glucose and 2 times a blend of sugars mimicking the composition of molasses. Biomass production on molasses was also higher than on the corresponding sugar blend, suggesting that non-sugar nutrients in molasses support growth. Similarly, when grape juice was used as a growth substrate, higher growth was obtained than on the corresponding sugar mixture [320].

Crude and inexpensive sources of carotenoid precursors could be used to enhance yield of astaxanthin. Alfalfa residual juice (ARJ), expected to have carotenoid precursors was studied as a growth substrate for *P. rhodozyma*. However, ARJ levels above 1.25% decreased pigmentation and at 3.7% completely repressed carotenogenesis [327]. Growth was not significantly affected by ARJ. Saponin, a terpenoid-glycoside, was found to be responsible for inhibition of carotenoid formation [328]. Carotenoid depression was reversed by polysaturated fatty acids and sterols, but not by the addition of media components.

Attempts have been made to increase production of carotenoids in *P. rhodozyma* and other fungi by feeding of terpenoids and other chemicals such as phenols. Although β-ionone stimulates carotenogenesis in *Phycomyces blakesleeanus* [329] and *Blakeslea trispora* [330], it inhibited astaxanthin formation in *P. rhodozyma*. β-Ionone caused an accumulation of β-carotene and a depletion of xanthophylls. Feeding of the monoterpene α-pinene to a mutant of *P. rhodozyma* increased the total pigment content from 1652 to 2201 $\mu g\,g^{-1}$, but there was a sharp decrease in growth rate at the effective concentration [331]. Other monoterpenes were too toxic to be used as supplements. Meyer and du Preez [332] investigated whether acetic acid would increase pigmentation with the assumption that it could serve as a precursor to acetyl-CoA. However, low concentrations of acetic acid (2 $g\,l^{-1}$) decreased the growth rate and astaxanthin production by *P. rhodozyma* on glucose. Titrating the culture with acetic acid to maintain pH control increased the biomass and astaxanthin. In *Phycomyces*, a variety of compounds are known to stimulate synthesis (mostly terpenoids and phenols) and recently an unidentified fungal metabolite produced by an *Apsergillus* sp was stimulatory [333]. Potential stimulants that work independently or synergistically would be worthy of investigation for industrial *P. rhodozyma* production.

Considerable progress has been made by several bio-industry companies in the development of the yeast *Phaffia rhodozyma* as a source of pigmentation and other nutrients for aquaculture. In Milwaukee, Wisconsin, Universal Bioventures has commercialized a yeast product that producers at least 4000 $\mu g\,g^{-1}$ of astaxanthin in reactor culture. Other companies also have developed processes for production of astaxanthin by yeast. Most of the strains with enhanced pigmentation have been obtained primarily by the use of chemical mutagenesis and visual screening coupled with the use of inhibitors of the isoprenoid pathoway. However, unwanted effects are often associated with these methods including genetic instability in overproducing strains, lack of reproducibility in industrial fermentations, and decreased growth of mutants compared to the wild-type.

There are presently several constraints in developing a commercial astaxanthin process utilizing *P. rhodozyma* including: (a) a scarcity of basic genetic methods for understanding carotenogenesis and construction of stable hyper-producing strains, (b) a lack of understanding of cellular metabolism in relation to astaxanthin synthesis, (c) complexities in bioprocess control and monitoring, (d) lack of technologies to detect and isolate astaxanthin hyperproducing mutants, (e) and the intracellular nature of astaxanthin. Progress is being made in the development of basic genetic methods and in optimization of process conditions. The isolation and characterization of the genes and enzymes responsible for xanthophyll formation in *P. rhodozyma* would be a significant advance but this information is not yet available. For strain development, it is possible that automated technologies could be optimized for detection of and isolation of carotenogenic mutants including FACS possibly combined with positive genetic selections [334] and physiological measurements [335]. Confocal redox and fluorescence imaging of single cells [312, 336], and digital imaging spectroscopy of cells and colonies may be valuable for detecting pigment mutants [337, 338]. A bottleneck in obtaining the highest production of astaxanthin is its intracellular compartmentalization. An understanding of its site of synthesis, compartmentalization, and transport throughout the cell could possibly lead to new technologies such as the potential use of permeabilized cells for end product removal and accumulation in an extractive phase [339].

9 Future Prospects

The production of carotenoids by microbial synthesis represents a classic example of competition between biological and chemical processes. It is unlikely that biosynthesis will be a significant source of the simpler carotenoids that are produced by highly advanced chemical syntheses. Microorganisms have potential for industrial production of carotenoids with complex structures including many xanthophylls that exist in nature as configurational isomers. The phylogeneic distribution of carotenoids in microorganisms confirms that specific groups have evolved the ability to synthesize carotenoids of medical and industrial interest. This analysis also indicates that certain ecological niches possessing carotenogenic organisms are unexplored and potentially valuable organisms remain to be isolated.

Of the various functions of carotenoids, a unifying mechanism is the ability of this class of compounds to quench $^1O_2$ and possibly other reactive forms of oxygen. Knowledge of the biological roles of carotenoids in humans and lower life is advancing at a rapid rate which should provide an impetus for increased intensity of research. It is refreshing to realize that besides providing aesthetic beauty and photoprotection in nature, carotenoids may be quite important in increasing the health and longevity of humans and animals by preventing chronic diseases.

Appendix

Trivial Name	Systematic Name
Aleuriaxanthin	(2R)-1',16'-Didehydro-1',2'-dihyro-β,ψ-caroten-2'-ol
Alloxanthin	(3R,3'R)-7,8,7',8'-Tetradehydro-β,β-carotene-3,3'-diol
Antheraxanthin	5,6-Epoxy-5,6-dihydro-β,β-carotene-3,3'-diol
β-Apo-8'-carotenoic acid	8'-Apo-β-caroten-8'-oic acid
Astaxanthin	3,3'-Dihydroxy-β,β-carotene-4,4'-dione
Bacterioruberin	(2S,2'S)-2,2'-Bis(3-hydroxy-3-mehylbutyl)-3,4,3',4'-tetradehydro-1,2,1',2'-tetrahydro-ψ,ψ-carotene-1,1'-diol
Bixin	Methyl hydrogen 9'-cis-6,6'-diapocarotene-6,6'-dioate
Caloxanthin	(2R,3R,3'R)-β,β-Carotene-2,3,3'-triol
Canthaxanthin	β,β-Carotene-4,4'-dione
Capsanthin	(3R,3'S,5'R)-3,3'-Dihydorxy-β,x-caroten-6'-one
Capsorubin	(3S,5R,3'S,5'R)-3,3'-Dihydroxy-x,x-carotene-6,6'-dione
α-Carotene	(6'R)-β,ε-Carotene
β-Carotene	β,β-Carotene
γ-Carotene	β,ψ-Carotene
ζ-Carotene	7,8,7',8'-Tetrahydro-ψ,ψ-carotene
Chlorobactene	φ,ψ-Carotene
Chloroxanthin	1,2,7',8'-Tetrhydro-ψ,ψ-caroten-1-ol
Citranaxanthin	5',6'-Dihydro-5'-apo-18'-or-β-caroten-6'-one
Crocin	Digentibiosyl-8,8'-diapocarotene-8,8'-dioate
Crocoxanhin	(3R,6'R)-7,8-Didehydro-β,ε-caroten-3-ol
Cryptoxanthin	(3R)-β,β-Caroten-3-ol
Decaprenoxanthin	(2R,6R,2'R,6'R)-2'-Bis-(4-hydroxy-3-methyl-2-butenyl)-ε,ε-carotene
Deoxyflexixanthin	1'-Hydroxy-3',4'-didehydro-1',2'-dihydro-β,ψ-caroten-4-one
Diadinoxanthin	(3S,5'R,6S,3'R)-5,6-Epoxy-7',8'-didehydro-5,6-dihydro-β,β-carotene-3,3'-diol
4',4'-Diaponeurosporene	7,8-Dihydro-4,4'-diapocarotene
4,4'-Diaponeurospren-4-oic acid	7',8'-Dihydro-4,4'-diapocaroten-4-oic acid
4,4'-Diapophytoene	7,8,11,12,7',8'11',12'-Octahydro-4,4'-diapocarotene
Diatoxanthin	(3R,3'R)-7,8-Didehydro-β,β-carotene-3,3'-diol
Echinenone	β,β-Carotene-4-one
Eutreptiellanone	(3S,5R,6S)-3,6-Epoxy-3',4',7'8'-tetradehydro-5,6-dihydro-β,β-carotene-4-one
Flexixanthin	(3S)-3,1'-Dihydro-3',4'-didehydro-1',2'-dihydro-β,ψ-caroten-4-one
Fucoxanthin	(3S,5R,6S,3'S,5'R,6'R)-5,6-Epoxy-3,3',5'-trihydroxy-6',7'-didehydro-5,6,7,8,5',6'-hexahydro-β,β-caroten-8-one-3'-acetate
Heteroxanthin	(3S,5S,6S,3'R)-7',8'-Didehydro-5,6-dihydro-β,β-carotene-3,5,6,3'-tetrol
19'-Hexanoyloxyfucoxanthin	(3S,5R,6S,3'S,5'R,6'S)-5,6-Epoxy-3,3',5',19'-tetrahydroxy-6',7'-didehydro-5,6,7,8,5',6'-hexahydro-β,β-carotene-8-oe 3'-acetate 19'-hexanoate
Hydroxyspheroidenone	1'-Methoxy-3',4'-didehydro-1,2,7,8,1',2'-hexahydro-ψ,ψ-caroten-1-ol
Isorenieratene	φ,φ-Carotene
β-Isorenieratene	β,φ-Carotene
Leoprotene	see Isorenieratene
Loroxanthin	(3R,3'R,6'R)-β,ε-carotene-3,19,3'-triol
Lutein	(3R,3'R,6'R)-β,ε-Carotene-3,3'-diol
Lycopenal	13-cis-ψ,ψ-Carotene-20-al
Lycopene	ψ,ψ-Carotene
Lycopenol	ψ,ψ-Carotene-16-ol
Lycoxanthin	see Lycopenol
Mutachrome	5,8-Epoxy-5,8-dihydro-β,β-carotene

Appendix (*continued*)

Monadoxanthin	7,8-Didehydro-β,ε-carotene-3,3'-diol
Myxobactone	1'-Glucosyloxy-3',4'-didehydro-1',2'-dihydro-β,ψ-carotene-4-one
Myxoxanthophyll	2'-(β-L-Rhamnopyranosyloxy)-3,4'-didehydro-1',2'-dihydro-β,ψ-carotene-3,1':diol
Neoxanthin	(3S,5R,6R,3'S,5'R,6'S)-5',6'-Epoxy-6,7-didehydro-5,6,5',6'-tetrahydro-β,β-carotene-3,5,3'-triol
Neurosporene	7,8-Dihydro-ψ,ψ-carotene
Neurosporoxanthin	4'-Apo-β-carotene-4'-oic acid
Nonaprenoxanthin	2-(4-Hydroxy-3-methyl-2-butenyl)-7',8',11',12'-tetrahydro-ε,ψ-carotene
Norbixin	6,6'-Diapocarotene-6,6'-dioic acid
Nostoxanthin	(2R,3R,2'R,3'R)-β,β-Carotene-2,3,2',3'-tetrol
Okenone	1-Methoxy-1',2'-dihydro-χ,ψ-caroten-4'-one
Oscillaxanthin	2R,2'R)-2,2'-Bis(β-L-rhamnopyransyloxy)-3,4,3',4'-tetradehydro-1,2,1',2'-tetrahydro-ψ,ψ-carotene-1,1'-diol
Peridinin	5',6'-Epoxy-3,5,3'-trihydroxy-6,7-didehydro-5,6,5',6'-tetrahydro-10,11,20-trinor-β,β-caroten-19',11'-olide-3-acetate
Phillipsiaxanthin	1,1'-Dihydrxy-3,4,3',4'-tetradehydro-1,2,1',2'-tetrahydro-ψ,ψ-carotene-2,2'-dione
Phleixanthophyll	(2'S)-1'-(β-D-Glucopyranosyloxy)-3',4'-didehydro-1',2'-dihydro-β,ψ-caroten-2'-ol
Phytoene	15-cis-7,8,11,12,7',8',11',12'-Octohydro-ψ,ψ-carotene
Phytofluene	15-cis–7,8,11,12,7',8'-Hexahydro-ψ,ψ-carotene
Plectaniaxanthin	(2'R)-3',4'-Didehydro-1',2'-dihydro-β,ψ-carotene-1',2'-diol
Prasinoxanthin	(3'R,6'R)-3,6,3'-Trihydroxy-7,8-dihydro-γ,ε-caroten-8-one
Prephyoene pyrophosphate	7,8,11,12,13,14,15,7',8',11',12',15'-Dodecahydro-13,15': 14,15'-bicyclo-15,15'-seco-ψ,ψ-carotene-15-ol 15 pyrophosphate
Rhodopin	1,2-Dihydro-ψ,ψ-carotene-1-ol
Rhodopinal	13-cis-1-Hydroxy-1,2-dihydro-ψ,ψ-caroten-20-al
Rhodoxanthin	4',5'-Didehydro-4,5'-retro-β,β-carotene-3,3'-dione
Saproxanthin	3',4'-Didehydro-1',2'-dihydro-β,ψ-carotene-3,1'-diol
Sarcinaxanthin	2'-(4-Hydroxy-3-methyl-2'-butenyl)-2-(3-methyl-2-butenyl)-ε,ε-caroten-18-ol
Siphonaxanthin	3,19,3'-Trihydroxy-7,8-dihydro-β,ε-carotene-8-one
Siphonein	(3R,3'R,6'R)-3,19,3'-T rihydroxy-7,8-dihydro-β,ε-carotene-8-one-19-laurate
Spheroidene	1-Methoxy-3,4,-didehydro-1,2,7',8'-tetrahydro-ψ,ψ-carotene
Spirilloxanthin	1,1'-Dimethoxy-3,4,3',4'-tetradehydro-1,2,1',2'-tetrahydro-ψ,ψ-carotene
Torularhodin	3',4'-Didehydro-β,ψ,-carotene-16'-oic acid
Torulene	3',4'-Didehydro-β,ψ-carotene
Violaxanthin	(3S,5R,6S,3'S,5'R6'S)-5,6,5',6'-Diepoxy-5,6,5',6'-tetrahydro-β,β-carotene-3,3'-diol
β-Zeacarotene	7',8'-Dihydro-β,ψ-carotene
Zeaxanthin	(3R,3'R)-β,β-Carotene-3,3'-diol

10 References

1. Griffiths M, Sistrom WR, Cohen-Bazire G, Stanier RY (1955) Nature (London) 176:1211
2. Cohen-Bazire G, Stanier RY (1958) Nature (London) 181:250
3. Siefermann-Harms D (1987) Physiol Plant 69:561

4. Cavalier-Smith T (1992) Ciba Foundation Symp 171:64
5. Pierson BK, Olson JM (1989) In: Cohen Y, Rosenberg E (ed) Microbial mats:Physiological ecology of benthic microbial communities. Americal Society for Microbiology, Washington DC, p 402
6. Olson JM, Pierson BK (1987) Int Rev Cytol 108:209
7. Cavalier-Smith T (1993) Microbiol Rev 57:953
8. Lewin RA (1993) Origins of plastids. Chapman and Hall, New York
9. Stanier RY (1970) Twentieth symposium of the society for general microbiology. Cambridge University Press, Cambridge, p 1
10. Whatley JM (1981) New Phytol 87:233
11. Pfander H (1992) Meth Enzymol 213:3
12. IUPAC Commission of Biochemical Nomenclature (1971) In: Isler O (ed), Carotenoids, Birkhauser, Basel, p. 851
13. IUPAC Commission on Biochemical Nomenclature (1975) Biochemistry 14:1803
14. Straub O (ed) (1987) Key to Carotenoids, 2nd edn. Birkhauser, Basel
15. Woese CR, Fox GE (1977) Proc Natl Acad Sci USA 74:5088
16. Woese CR (1987) Microbiol Rev 51:221
17. Iwabe N, Kuma K-I, Hasegawa M, Osawa S, Miyata T (1989) Proc Natl Acad Sci USA 86:9355
18. Cavalier-Smith T (1993) Microbiol Rev 57:953
19. Bloch K (1991) In: Vance DE, Vance JE (ed) Biochemistry of lipids, lipoproteins and membranes. Elsevier, Amsterdam (New Comprehensive Biochemistry, vol 20), p 363
20. Cavalier-Smith T (1990) Rev Micropaleontol 33:145
21. Cavalier-Smith T (1987) Cold Spring Harbor Symp Quant Biol 52:805
22. Stanier RY, Pfennig N, Truper HG (1981) In: Starr MP, Stolp H, Truper HG, Balows A, Schlegel HG (eds) The prokaryotes. Springer, Berlin Heidelberg New York, p 197
23. Pfennig N, Truper HG (1989) In: Staley JT, Bryant MP, Pfennig N, Holt JG (eds) Bergey's manual of systematic bacteriology, vol 3. Williams and Wilkins, Baltimore, p 1635
24. Imhoff JF (1992) In: Mann NH, Carr NG (ed) Photosynthetic prokaryotes, Plenum, New York, p 53
25. Walker JCG, Klein C, Schidlowski M, Schopf JW, Stevenson DJ (1983) In: Schopf JW (ed), Earth's earliest biosphere, Princeton Univ Press, New Jersey, p 260
26. Walsh MM, Lowe DR (1985) Nature (London) 314:530
27. Gest H, Favinger JL (1983) Arch Microbiol 136:11
28. Kobayashi M, Van de Meent EJ, Erkelens C, Amesz J, Ikegami I, Watanabe T (1991) Biochim Biophys Acta 1057:89
29. Brock TD, Madigan MT, Martinko JM, Parker J (1994) Biology of Microorganisms, 7th edn. Prentice Hall, New Jersey, p 722
30. Schmidt K (1978) In: Clayton RK, Sistrom WR (ed) The photosynthetic bacteria Plenum, New York, p 729
31. Pfennig N (1989) In: Staley JT, Bryant MP, Pfennig N, Holt JG (eds) Bergey's manual of systematic bacteriology, vol. 3. Williams and Wilkins, Baltimore, p 1682
32. Castenholz RW (1989) In: Staley JT, Bryant MP, Pfennig N, Holt JG (eds) Bergey's manual of systematic bacteriology, vol. 3. Williams and Wilkins, Baltimore, p. 1698
33. Perry JJ (1992) In: Balows A, Truper HG, Dworkin M, Harder W, Schleifer K-H (eds) The prokaryotes. Springer, Berlin Heidelberg New York, p 3777
34. Walker JCG, Klein C, Schidlowski M, Schopf JW, Stevenson DJ (1983) In: Schopf JW (ed) Earth's earlist bioshpere, Princeton University Press, New Jersey, p 260
35. Lewin RA, Withers, NW (1975) Nature (London) 256:735
36. Post AF, Bullerjahn GS (1994) FEMS Microbiol Rev 13:393
37. Foss O, Lewin RA, Liaaen-Jensen S (1987) Phycologia 26:142
38. Whithers, NW, Alberte RS, Lewin RA, Thornber JP, Britton G, and Goodwin TW (1978) Proc Natl Acad Sci USA 75:2301
39. Chisolm SW, Olson RJ, Zettler ER, Goericke R, Waterbury JB, Welschmeyer NA (1988) Nature (London) 334:340
40. Lewin RA (1989) In: Staley JT, Bryant MP, Pfennig N, Holt JG (eds) Bergey's manual of systematic bacteriology, vol. 3. Williams & Wilkins, Baltimore, p 1800
41. Gray MW and Doolittle WF (1982) Microbiol Rev 46:1
42. Sagan L (1967) J Theor Biol 14:225
43. Margulis L (1981) Symbiosis in cell evolution. W.H. Freeman and Company, San Francisco

44. Bonen L, Doolittle WF (1975) Proc Natl Acad Sci USA 72:2310
45. Mereschowsky C (1905) Biol Zentralbl 25:593
46. Martin W, Somerville CC, Loiseaux-de Goer S (1992) J Mol Evol 35:385
47. Gray MW (1989) Trends Genet 5:294
48. Stanier RY, Cohen-Bazire G (1957) In: Williams REO, Spicer CC (eds) Microbial ecology, Cambridge University Press, U.K., p 56.
49. Imhoff JF (1992) In: Mann NH, Carr NG (eds) Photosynthetic prokaryotes, Plenum, New York, p 53
50. Kamekura M (1993) In: Vreeland RH, Hochstein LI (eds) The biology of halophilic bacteria, CRC Press, Boca Raton, Floride, p 150
51. Dundas ID, Larsen H (1962) Arch Mikrobiol 44:233
52. Oesterhelt D, Stoeckenius (1971) Nature 233:149
53. Balows A, Truper HG, Dworkin M, Harder W, Schleifer K-H (eds) (1992) The prokaryotes, 2nd edn. Springer, Berlin Heidelberg, New York
54. Goodwin TW (1980) The biochemistry of the carotenoids, 2nd edn. Chapman and Hall, London, p 291
55. Ratledge C, Wilkinson SG (1988) Microbial lipids. Academic Press, London, pp 55, 117, 299
56. Beyer P, Kleinig H, Englert G, Meister W, Noack K (1979) Helv Chim Acta 62:2551
57. Kleinig H, Broughton WJ (1982) Arch Microbiol 133:164
58. Jenkins CL, Andrewes AG, McQuade, Starr MP (1979) Curr Microbiol 3:1
59. Mancock IC, Williams KM (1986) J Gen Microbiol 132:599
60. Dobereiner J (1992) In: Balows A, Truper HG, Dworkin M, Harder W, Schleifer K-H (eds) The prokaryotes, 2nd edn. Springer, Berlin Heidelberg, New York, p 2236
61. Thomashow LS, Rittenberg SC (1985) J Bacteriol 163:1047
62. Starr MP (1981) In: Starr MP, Stolp H, Truper A, Balows A, Schlegel (eds) The prokaryotes, vol II. Springer, Berlin, Heidelberg New York, p 1260
63. Starr MP, Jenkins CL, Bussey LB, Andrewes, AG (1977) Arch Mikrobiol 113:1
64. Shiba T (1992) in: Balows A, Truper HG, Dworkin M, Harder W, Schleifer K-H (ed) The Prokaryotes, second edition. Springer, Berlin, Heidelberg New York, p 2156
65. Weeks OB, Andrewes AG, Brown, BO, Weedon BCL (1969) Nature (London) 224:879
66. Personal communications
67. Reichenbach H (1992) in: Balows A, Truper HG, Dworkin M, Harder W, Schleifer K-H (ed) The Prokaryotes, second edition. Springer, Berlin Heidelberg New York, p 3631
68. Reichenbach H, Kleinig H (1971) Arch Mikrobiol 76:364
69. Reichenbach H, Dworkin M (1992) in; Balows A, Truper HG, Dworkin M, Harder W, Schleifer K-H (ed) The Prokaryotes, second edition. Springer-Verlag, New York, p 3416
70. Reichenbach H (1992) in: Balows A, Truper HG, Dworkin M, Harder W, Schleifer K-H (ed) The Prokaryotes, second edition. Springer-Verlag, New York, p 3676
71. Aasen AJ, Liaaen-Jensen S (1966) Acta Chem Scand 20:2322
72. Nelis HJ, De Leenheer AP (1989) In: Vandamme EJ (ed), Biotechnology of vitamins, pigments and growth factors. Elsevier, London, p 43
73. Goodwin TW (1972) Prog Indust Microbiol 11:29–88
74. Arpin N, Norgard S, Francis GW, Liaaen–Jensen S (1973) Acta Chem Scand 27:2321
75. Hertzberg S, Liaaen-Jensen S (1977) Acta Chem Scand Ser B 31:215
76. Swift IE, Milborrow BV (1981) J Biol Chem 256:11607
77. Nelis HJ, de Leenheer AP (1989) Appl Environ Microbiol 55:2505
78. Iizuka H, Nishimura Y (1969) J Gen Appl Microbiol 15:127
79. Taylor RF, Davies BH (1973) Biochem Soc Trans 1:1091
80. Taylor RF, Davies BH (1975) Proc 4th Symp Carotenoid, p. 66
81. Taylor RF, Davies BH (1976) J Gen Microbiol 92:325
82. Taylor RF, Davies BH (1976) Biochem J 153:233
83. Clejan S, Krulwich TA, Mondrus KR, Seto-Young D (1986) J Bacteriol 168:334
84. Thirkell D, Summerfield M (1980) Ant van Leeuwenhoek 46:51
85. Carbonneau MA, Melin AM, Perromat A, Clerc M (1989) Arch Biochem Biophys 275:244
86. Ray PH, White DC, Brock TD (1971) J Bacteriol 108:227
87. Rowan KS (1989) Photosynthetic pigments of algae. Cambridge University Press, Cambridge, p 112
88. Goodwin T, Britton GW (1988) In: Goodwin TW (ed), Plant pigments. Academic Press, London, p 61
89. Goodwin TW (1992) Meth Enzymol 213:167

 90. Gibbs SP (1990) In: Wiessner W, Robinson DG, Starr RC (eds). Experimental phycology, vol 1. Springer, Berlin Heidelberg New York, p 145
 91. Liaaen-Jensen S (1977) In Marine natural products: NATO special program panel on marine sciences. Plenum, New York, p 239
 92. Cavalier-Smith T (1993) In: Lewin RA (ed), Origins of plastids. Chapman and Hall, New York, p 291
 93. Bonen L, and Doolittle WF (1975) Proc Natl Acad Sci USA 72:2310
 94. Tangen K, Bjornland T (1981) J Plankton Res 3:389
 95. Ragan MA, Chapman DJ (1978) A biochemical phylogeny of the protists. Academic Press, New York.
 96. Bjornland T (1990) In: Krinsky NI, Mathews-Roth MM, Taylor RF (eds), Carotenoids. Chemistry and biology, Plenum, New York, p 21
 97. Liaaen-Jensen S, Andrewes AG (1972) Annu Rev Microbiol 26:225
 98. Ling HU, Seppelt, RD (1993) Eur J Phycol 28:77
 99. Ling HU, Seppelt, RD (1993) Eur J Phycol 28: 77
100. Viala G (1966) CR Hebd Seances Acad Sci 263:1383
101. Hagen C, Braune W, Bjorn LO (1994) J Phycol 30:241
102. Yong, YYR, Lee Y-K (1991) Phycologia 30:257
103. Hagen C, Braune W, Bjorn LO (1994) J Phycol 30:241
104. Ben-Amotz A, Shaish A, Avron M (1989) Plant Physiol 91:1040
105. Giovannoni SJ, Turner S, Olsen GJ, Barns S, Lane DJ, Pace NR (1988) J Bacteriol 170:3584
106. Gray MW 1992 Int Rev Cytol 141:233
107. Martin W Somerville CC, de Goer SL (1992) J Mol Evol 35:385
108. Reith M, Munholland J (1993) The Plant Cell 5:465
109. Michalowski CB, Loffelhardt W, Bohnert HJ (1991) J Biol Chem 266:11866
110. Bauldauf SL, Palmer, JD (1993) Proc Natl Acad Sci USA 90:11558
111. Wainright PO, Hinkle G, Sogin ML, Stickel SK (1993) Science 260:340
112. Kreger van Rij NJW (1984) The yeasts 3rd edn. Elsevier, Amsterdam.
113. Goodwin TW (1980) The biochemistry of the carotenoids, 2nd edn, vol I. Plants. Chapman and Hall, London, p 257
114. Phaff HJ, Miller MW, Yoneyama M, Soneda M (1972) Proc IV IFS: Ferment Technol Today, Kyoto, Society of Fermentaion Technology, Osaka, p 759
115. Miller MW, Yoneyama M, Soneda M (1976) Int J Syst Bacteriol 26:286
116. Andrewes AG, Phaff HJ, Starr MP (1976) Phytochem 15:1003
117. Andrews AG, Starr MP (1977) Phytochem 15:1009
118. Cerda-Olmedo E, Avalos J (1994) Prog Lipid Res 33: 185
119. Cavalier-Smith T (1987) In: Rayner ADM, Brasier CM, Moore D (eds), Evolutionary biology of the fungi (Symp Br Mycol Soc 13), Cambridge University Press, Cambridge, p 339
120. Armstrong GA, Hundle BS, Hearst JE (1993) Meth Enzymol 214:297
121. Wainright PO, Hinkle G, Sogin ML, Stickel SK (1993) Science 260: 340
122. Sleigh M (1989) Protozoa and other flagellates. Edward Arnold, London
123. Dembitsky, V.M. (1992) Prog Lipid Res 31:373
124. Ames BN, Shigenaga MK, Hagen TM (1993) Proc Natl Acad Sci USA 90:7915
125. Peto R, Doll R, Buckley JD, Sporn MB (1981) Nature (London) 290: 201
126. Hirayama T (1979) Nutr Cancer 1:67
127. Stadtman ER 1992 Science 257:1220
128. Sohal RS, Agarwal S, Dubey A, Orr WC (1993) Proc Natl Acad Sci USA 90:7255
129. Krinsky NI (1977) Trends Biochem Sci 2:35
130. Wasserman HH, Murray RW (eds) (1979) Singlet oxygen. Academic Press, New York
131. Adam W, Cilento G (1982) Chemical and biological generation of electronically excited states. Academic Press, New York
132. Krinsky NI (1974) Science 186:363
133. Epe B (1991) Chem-Biol Interact 67:149
134. Piette J (1991) J Photochem Photobiol B, Biol 11:241
135. Sies H, Menck CFM (1992) Mutat Res 275:367
136. Davies KJA (1987) J Biol Chem 262:9895
137. Wagner JR, Motchnik PA, Stocker R, Sies H, Ames BN (1993) J Biol Chem 268:18502
138. van der Vliet A, Bast A (1992) Free Rad Biol Med 12:499
139. Foote CS, Denny RW (1968) J Am Chem Soc 90:6233

140. Palozza P, Krinsky NI (1992) Meth Enzymol 213:403
141. Hirayama O, Nakamura K, Hamada S, Koboyasi Y (1994) Lipids 29:149
142. Di Mascio P, Sundquist AR, Devasagayam PA, Sies H (1992) Meth Enzymol 213:429
143. Sundquist AR, Brivibak K, Sies H (1994) Meth Enzymol 234:384
144. Satamaria L, Bianchi A, Arnaboldi A, Andreoni L, Bermond P (1983) Experientia 39:1043
145. Gerster H (1993) Internat J Vit Nutr Res 63:93
146. Seifter E, Rettura G, Padwaer J, Levenson SM (1987) J Natl Cancer Inst 68:835
147. Murakoshi M, Nishino H, Satomi S, Takayusu J, Hasegasa T, Tokuda H, Iwashima A, Okuzumi J, Okabe H, Kitano H, Iwasaki R (1992) Cancer Research 52:6583
148. Tanaka T, Morishita Y, Suzui M, Kojima T, Okumura A, Mori H (1994) Carcinogenesis 15:15
149. Jyonouchi H, Zhang L, Gross M, Tomita Y (1994) Nutr Cancer 21:47
150. Jyonouchi H, Zhang L, Tomita Y (1993) Nutr Cancer 19:269
151. Zigman S (1993) Photochem Photobiol 57:1060
152. Balasubramanian D, Du X, Zigler JS Jr (1990) Photochem. Photobiol 52:761
153. Packer L. (1993) Ann N Y Acad Sci 691:48
154. Bendich A (1993) Biological functions of dietary carotenoids. Ann N Y Acad Sci 691:61
155. Olson JA (1993) Ann N Y Acad Sci 691:156
156. Quinones MA, Zeiger E (1994) Science 264:558
157. Parry AD, Horgan R (1992) Planta 187:185
158. Saranak J, Foster KW (1994) J Exp botany 45:505
159. Davies BW, Davies BH (1986) Biochem Soc Trans 14:952
160. Gudas LJ (1994) J Biol Chem 269:15399
161. Miki W (1991) Pure Appl Chem 63:141
162. Childress JJ, Fisher CR (1992) Oceangr Mar Annu Rev 30:337
163. Rau W (1988) In: Goodwin TW (ed) Plant pigments. Academic Press, London, p 231
164. Masood A, Dogra JVV, Jha AK (1994) Lett Appl Microbiol 18:184
165. Krinsky NI (1968) In: Giese AC (ed) Photophysiology, vol III, Academic Press, New York, p 123
166. Rohmer M, Bouvier P, Ourisson G (1979) Proc Natl Acad Sci USA 76:847
167. Lazrak T, Wolff G, Albrecht A-M, Nakatani Y, Ourisson G, Kates M (1988) Biochim Biophys Acta 939:160
168. Chamberlain NR, Mehrtens BG, Xiong Z, Kapral FA, Boardman JL, Rearick JI (1991) Infect Immun 59:4332
169. Bramley PM, Bird CR, Schuch W (1993) In:Grierson D (ed) Biosynthesis and manipulation of plant products, vol 3, p 139
170. Sandmann G (1991) Physiologia Plantar 83:186
171. Bloch KE (1983) Crit Rev Biochem 14:47
172. Basson ME, Thorsness M, Rine J (1986) Proc Natl Acad Sci USA 83:5563
173. Enjuto M, Balcells L, Campos N, Caelles C, Arro M, Boronat A (1994) Proc Natl Acad Sci USA 91:927
174. Wright R, Basson M, D'Ari L, Rine J (1988) J Cell Biol 107:101
175. Hampton RY, Rine J (1994) J Cell Biol 125:299
176. Basson ME, Thorsness M, Finer RM, Stroud RM, Rine J (1988) Mol Cell Biol 8:797
177. Young J, Kane LP, Exley M, Wileman T (1993) J Biol Chem 268:19810
178. Casey WM, Keesler GA, Parks LW (1992) J Bacteriol 174:7283
179. Bejarano ER, Cerda'-Olmedo E (1992) FEBS Lett 306:209
180. De la Guardia MD, Aragon CMG, Murillo FJ, Cerda-Olmedo E (1971) Proc Natl Acad Sci USA 68:2012
181. Aragon CMG, Murillo FJ, De la Guardia MD, Cerda-Olmedo E (1976) Eur J Biochem 63:71
182. Candau R, Bejarano ER, Cerda-Olmedo E (1991) Proc Natl Acad Sci USA 88:4936
183. McDermott JCB, Brown DJ, Britton G, Goodwin TW (1974) Biochem J 144:231.
184. Coolbear T, Threlfall DR (1989) In: Ratledge C, Wilkinson SG (eds), Microbial lipids, vol 2. Academic Press, London, p 115
185. Santos F, Mesquita JF (1984) Cytologia 49:215
186. Hagen Ch, Braune W, Greulich F (1993) J Photochem Photobiol B: Biol 20:153
187. McCully EK, Bracker CE (1972) J Bacteriol 109:922
188. Damsky CH (1976) J Cell Biol 71:123
189. Wanner G, Formanek H, Theimer RR (1981) Planta 151:109
190. Marano MR, Serra EC, Orellano EG, Carillo N (1993) Plant Sci 94:1

191. Veenhuis M, Harder W (1991) In: Rose AH, Harrison JS (eds) The yeasts, 2nd edn, vol 4. Academic Press, London, p 601
192. Patori GM, del Rio LA (1994) Planta 193:385
193. Munkres KD (1990) Rev Biol Res Aging 4:29
194. Stamellos KD, Shackelford JE, Tanaka RD, Krisans SK (1992) J Biol Chem 267:5560
195. Ericsson J, Appelkvist E-L, Thelin A, Chojnacki T, Dallner G (1992) J Biol Chem 267:18708
196. Armstrong GA, Alberti M, Leach F, Hearst JE (1989) Mol Gen Genet 216:254
197. Coomber SA, Chaudri M, Connor A, Britton G, Hunter CN (1990) Mol Microbiol 4:977
198. Lee L-Y, Liu ST (1991) Mol Microbiol 5:275
199. Yo K-Y, Lai E-M, Lee L-Y, Lin T-P, Hung C-H, Chen C-L, Chang Y-S, Liu S-T (1994) Microbiology 140:331
200. Misawa N, Nakagawa M, Kobayashi K, Yamao S, Isawa Y, Nakamura K, Harashima K (1990) J Bacteriol 172:6704
201. Ruiz-Vazquez R, Fontes M, Murillo FJ (1993) Mol Microbiol 10:25
202. Houssaini-Iraqui M, Clavel-Seres S, Rastogi N, David HL (1993) Curr Microbiol 26:65
203. Tabata K, Ishida S, Nakahara T, Hoshino T (1994) FEBS Lett 341:251
204. Gantotti BV, Beer SV (1982) J Bacteriol 151:1627
205. Wellington CL, Bauer CE, Beatty JT (1992) Canad J Microbiol 38:20
206. Armstrong GA, Alberti M, Hearst JE (1990) Proc Natl Acad Sci USA 87:9975
207. Armstrong GA, Schmidt A, Sandmann G, Hearst JE (1990) J Biol Chem 265:8329
208. Pemberton JM, Harding CM (1987) Curr Microbiol 15:67
209. Hoshino T, Fujii R, Tadaatsu N (1994) J Ferm Bioeng 77:423
210. Armstrong GA, Hundle BS, Hearst JE (1992) Meth Enzymol 214:297
211. Sandmann G (1991) Physiol Plant 83:186
212. Pecker I, Chamowitz D, Linden H, Sandmann G, Hirschberg J (1992) Proc Natl Acad Sci USA 89:4962
213. Schmidhauser TJ, Lauter F-R, Schumacher M, Zhou W, Russo VEA, Yanofsky C (1994) J Biol Chem 269:12060
214. Goldie AH, Subden RE (1973) Biochem Genet 10:275
215. Schmidhauser TJ, Lauter FR, Russo VEA, Yanofsky C (1990) Mol Cell Biol 10:5064
216. Nelson MA, Morelli G, Carottoli A, Romano N, Macino G (1989) Mol Cell Biol 9:1271
217. Schmidhauser TJ, Lauter F-R, Schumacher M, Zhou W, Russo VEA, Yanofsky C (1994) J Biol Chem 269:12060
218. Harding RW, Turner RV (1981) Plant Physiol 68:745
219. Armstrong GA, Alberti M, Hearst JE (1990) Proc Natl Acad Sci USA 87:9975
220. Cohen-Bazire G, Sistrom WR, Stanter RY (1957) J Cell Comp Physiol 49:25–68
221. Drews G (1986) Trends Biochem Sci 11:255
222. Penfold RJ, Pemberton JM (1994) J Bacteriol 176:2869
223. Penfold RJ, Pembertyon JM (1991) Curr Microbiol 23:25
224. Hodgson DA, Murillo FJ (1993) In: Dworkin M, Kaiser D (eds) Myxobacteria II. Amer Soc Microbiol, Washington DC, p 157
225. Farr SB, Kogoma T (1991) Microbiol Rev 55:561
226. Burchard RP, Hendricks SB (1969) J Bacteriol 97:1165
227. Candau R, Bejarano ER, Cerda-Olmedo E (1991) Proc Natl Acad Sci USA 88:4936
228. Bejaron ER, Parra F, Murillo FJ, Cerda-Olmedo E (1988) Arch Microbiol 150:209
229. Salgado LM, Cerda-Olmedo E (1992) Curr Genet 21:67
230. Corrochano LM, Avalos J (1992) Exp Mycol 16:167
231. Golubev WI (1995) Yeast 11:101
232. Ludwig F (1891) Centralbl Bakteriol Parasitenk 10:10
233. Ludwig F (1896) Centralbl Bakteriol Parasitenk Abt II 2:337
234. Wieth T (1991) Development of a yeast product with high astaxanthin content. Thesis Akademiet for de tekniske videnskaber, Copenhagen Dk
235. Evans CT, Adam D, Wisdom RA (1990) European Patent Application #90311254.8 (15.10.90)
236. Girard P, Falconnier B, Bricout J, Vladescu B (1994) Appl Microbiol Biotechnol 41:183
237. Schroeder WA, Johnson EA (1994) Unpublished results.
238. Bauernfeind JC (1981) In: Bauernfeind JC (ed) Carotenoids as colorants and vitamin A precursors, Academic Press, New York, p 1
239. Klaui H, Bauernfeind JC (1981) In: Bauernfeind JC (ed) Carotenoids as colorants and vitamin A precursors, Academic Press, New York, p 47

240. Dean KL (ed) (1992) Industrial Bioprocessing 14:6. Technical Insights, Inc., Englewood, New Jersey.
241. Natural Marine Fisheries (1992) World salmon culture
242. Rosenberry B (1993) Fisheries statistics yearbook, World shrimp farming.
243. Sunde ML (1992) Poultry Sci 71:709
244. Nelis HJ, DeLeenher AP (1989) In: Vandamme EJ (ed) Biotechnology of vitamins, pigments and growth factors. Elsevier, London, p 43
245. Simpson KL, Katayama T, Chichester CO (1981) In: Bauernfeld JC (ed) Carotenoids as colorants and vitamin A precursors. Academic Press, London, p 463
246. Marusich WL, Bauernfeind JC (1981) In: Bauernfeld JC (ed) Carotenoids as colorants and vitamin A precursors. Academic Press, London, p 320
247. Ratcliff RG, Day EJ, Hill JE (1959) Poult. Sci. 38:1039
248. Nelis HJ, DeLeenher AP (1991) J. Appl Bacteriol. 70:181
249. Anon. (1962) Feeds Illus Nov: 24
250. Nelis HJ, DeLeenher AP (1991) J Appl Bacteriol 70:181
251. Theriault RJ (1965) Appl Microbiol 13:402
252. Ninet L, Renault J (1979) In: Peppler HJ, Perlman D, (eds) Microbiol Technology 2nd edn. Academic Press, New York, p 529
253. Orndorff S, Presentation at Soc Industr Microbiol Ann Mtg, San Diego, 1992
254. Luetke-Brinkhaus F (1993) Seminar Presented at the 93rd Annual Meeting of the American Society for Microbiology, 26 May 1993
255. Downs J, Harrison DEF (1974) J Appl Bacteriol 37:65
256. Haxo F (1950) Botanical Gazette 112:228
257. Czygan F-C (1968) Archiv fur Mikrobiol 61:81
258. Borowitzka LJ, Borowitzka MA (1989) In: Vandamme EJ (ed), Biotechnology of vitamins, pigments and growth factors. Elsevier, London, p. 15
259. Cera-Olmedo E (1989) In: Vandamme EJ (ed) Biotechnology of vitamins, pigments and growth factors. Elsevier, London, p 27
260. Ciegler A (1965) Adv Appl Microbiol 7:1
261. Shaish A, Ben-Amotz A, Avron M (1992) Meth Enzymol 213:439
262. Davies BH, Hsu W-J, Chichester CO (1965) Biochem J 94:26
263. Hata M, Hata M (1975) Tohoku J Agric Res 26:35
264. Bubrick P (1991) Bioresource Technol. 38:237
265. Neamtu G, Tamas V, Bodea C (1966) Rev Roum Biochim 3:305
266. Storebakken T (1988) Aquaculture 70:193
267. Goodwin TW, Jamikorn M (1953) Biochem J 57:376
268. Borowitzka MA, Huisman JM, Osborn A (1991) J Appl Phycol 3:295
269. Boussiba S, Fan L, Vonshak A (1992) Meth Enzymol 213:386
270. Kobayashi M, Kakizono T, Nagai S (1993) Appl Environ Microbiol 59:867
271. Kobayashi M, Kakizono T, Nagai S (1991) J Ferm Bioeng 71:335
272. Kobayashi M, Kakizono T, Nagai S (1993) Seibutsu-Kogaku 71:233
273. Lwoff M, Lwoff A (1930) C R Soc Biol Paris 105:454
274. Droop MR (1955) Nature (London) 175:42
275. Donkin P (1976) Phytochem 15:711
276. Renstrom B, Borch G, Skulberg OM, Liaaen-Jensen S (1981) Phytochem 20:2561
277. Golubev WI, Babjeva IP, Blagoodatskaya VM, Reshetova IS (1977) Mikrobiologiya 46:564
278. Yamada Y, Kawasaki H (1989) Agric Biol Chem 53:2845
279. Geuho E, Kurtzman CP, Peterson SW (1989) System Appl Microbiol 12:230
280. Boekhout T, Fonseca A, Sampaio J-P, Golubev WI (1993) Canad J Microbiol 39:276
281. Castillo A, Cifuentes V (1994) Curr Gen 26:364
282. Phrillinger H, Laaser G, Dorfler C, Ziegler K (1991) Sydowia 53:(in press)
283. Webster J (1980) Introduction to the Fungi, 2nd edn. Cambridge Univ Press, Cambridge, p 463
284. Swann EC, Taylor JW (1993) Mycologia 85:923
285. Bjerking B (1990) Quantitative and Qualitative investigation of the carotenoids of two cultivates of the yeast *Phaffia rhodozyma* for use in the cultivation of salmon (*Salmo salar*, L.) and rainbow trout (*Salmo gairdneri*, R.). University of Trondheim External report.
286. An G-H, Schuman DB, Johnson EA (1989) Appl Environ Microbiol 55:116
287. Lewis MJ, Ragot N, Berlant MC, Miranda M (1990) Appl Environ Microbiol 56:2944

288. Chun SB, Chin JE, Bai S, An G-H (1992) FEMS Microbiol Lett 93:221
289. Meyer PS, DuPreez JC, Kilian SG (1993) World J Microbiol Biotechnol 9:514
290. Schroeder WA, Johnson EA (1995) J Biol Chem 270:18374
291. Chang K-W (1990) Stimulation of carotenoid synthesis in *Phaffia rhodozyma* by metal ions and superoxide radicals. Thesis University of Wisconsin, Madison
292. Schroeder WA, Johnson EA (1993) J Gen Microbiol 139:907
293. Foote CS, Denny RW (1968) J Am Chem Soc 90:6233
294. Moan J (1990) J Photochem Photobiol B, Biol 6:343
295. Bradley DG, Min DB (1992) Crit Rev Food Sci Nutr 31:211
296. Hirayama O, Nakamura K, Hamada S, Kobayasi Y (1994) Lipids 29:149
297. DiMascio P, Sandquist AR, Devasagayam TPA, Sies H (1992) in: Packer L (ed) Meth Enzymol 213:429
298. Schroeder WA, Johnson EA (1995) J Ind Microbiol 14:502
299. Khan AU, Kasha M (1994) Proc Natl Acad Sci USA 91:12365
300. Lemke PA (1994) Mycologia 86:173
301. Erdman JL (1992) Molec Cell Biol 12:153
302. Tully M, Gilbert HJ (1985) Gene 36:235
303. Van Ooyen AJJ (1993) European Patent Appl 93202637.0, 10.09.93
304. Johnson EA, An G-H (1991) Crit Rev Biotechnol 11:297
305. Prevatt WD (1991) Spheroplast fusions of *Phaffia rhodozyma* cells. World Patent Application #9015322
306. Adrio JL, Veiga M, Casqueiro J, Lopez M, Fernandez C (1993) J Gen Appl Microbiol 39: 303
307. Villa TG (1994) Personal communication
308. Meyer PS, Wingfield BD, DuPreez JC (1994) Biotechnol Techniq 8:1
309. Varga J, Vagvolgyi C, Nagy A, Pfeiffer I, Ferenczy L (1995) Int J Syst Bact 45:173
310. Sensen CW, Heimann K, Melkonian M (1993) Eur J Phycol 28:93
311. Sakamoto K, Iida K, Koyano T, Asada Y, Furuya T (1994) Planta Med 60:253
312. An G-H, Bielich J, Auerbach R, Johnson EA (1991) Bio/Technol 9:70
313. An G-H (1991) Improved astaxanthin production from the red yeast *Phaffia rhodozyma*. Thesis University of Wisconsin, Madison
314. Kelly SE (1990) A design tool for the analysis of biochemical reaction networks utilizing stoichiometric structure. Thesis University of Wisconsin, Madison
315. Johnson EA, Lewis MJ (1979) J Gen Microbiol 115:173
316. Meyer PS, DuPreez JC (1994) Appl Microbiol Biotechnol 40:780
317. Fang TJ, Cheng Y-S (1993) J Ferm Bioeng 75:466
318. Poluyakh OV, Podoprigora OI, Eliseev SA, Ershov YV, Bykhovskii VY, Dmitrovskii AA (1991) Appl Biochem Microbiol 27:411
319. Prevatt WD, Dickson TD, Harris RL (1991) Novel strains of *Phaffia rhodozyma* containing high levels of astaxanthin. World Patent #8712855
320. Meyer PS, DuPreez JC (1994) World J Microbiol Biotechnol 10:178
321. Boveris A, Cadenas E (1975) FEBS Lett 54:311
322. An G-H, Johnson EA (1990) Antonie van Leeuwenhoek 57:191
323. Berthold DA, Siedow JN (1993) Plant Physiol 101:113
324. Krinsky NI (1971) In: Isler O (ed) Carotenoids, Birkhauser, Basel, p 669
325. Donaldson RP, Luster DG (1991) Plant Physiol 96:669
326. Haard NF (1988) Biotechnol Lett 10:609
327. Okagbue RN, Lewis MJ (1984) Appl. Microbiol. Biotechnol. 20:278
328. Okagbue RN, Lewis MJ (1984) Biotechnol. Lett. 6:247
329. Mackinnney G, Nakayama T, Chichester CO, biss CD (1953) J Am Chem Soc 75:236
330. Ninet L, Renaut J, Tissier R (1969) Biotechnol Bioeng 11:1195
331. Meyer PS, du Preez JC, van Dyk (1994) Biotechnol Lett 16:125
332. Meyer PS, du Preez JC (1993) Biotechnol Lett 15:919
333. Margalith PZ (1993) Appl Microbiol Biotechnol 38:664
334. Gross H-J, Verwer B, Houck D, Recktenwald D (1993) Cytometry 14:519
335. Skowronek P, Krummeck G, Haferkamp O, Rodel G (1990) Curr Genet 18:265
336. Masters BR (1994) Adv Molec Cell Biol 8:1
337. Yang MM, Youvan DC (1988) Bio/Technol 6:939
338. Arkin AP, Goldman ER, Robles SJ, Goddard CA, Coleman WJ, Yang MM, Youvan DC (1990) Bio/Technol 8:746
339. Brodelius P (1988) Appl Microbiol Biotechnol 27:561

Author Index Volume 53

Author Index Vols. 1-50 see Vol. 50

Subject Index